Lasers in Materials Processing

A BIBLIOGRAPHY OF A
DEVELOPING TECHNOLOGY

Alan Gomersall

Springer-Verlag Berlin Heidelberg GmbH

1986

British Library Cataloguing in Publication Data
Gomersall, Alan
  Lasers in materials processing: a bibliography of a developing technology
  1. Laser industry — Bibliography
  I. Title
  016.67042   Z5853.L3/

  ISBN 978-3-662-30196-8   ISBN 978-3-662-30194-4 (eBook)
  DOI 10.1007/978-3-662-30194-4

© 1986 Springer-Verlag Berlin Heidelberg
  Originally published by IFS (Publications) Ltd, 35-39 High Street, Kempston,
  Bedford MK42 7BT, UK and Springer-Verlag Berlin Heidelberg New York Tokyo in 1986

Phototypeset by Fleetlines Typesetters, Southend-on-Sea, Essex

Alan Gomersall, M.Phil., B.Sc.(Eng)., M.I.Inf.Sc.

Graduated as a mechanical engineer through London University and joined the English Electric Co. Ltd as an information officer. After six years lecturing in information science at Leeds Polytechnic he is now Head of Research Library, London Research Centre. Previous publications include: "Thesaurafacet: A Classification and Thesaurus for Science and Technology", "Traffic Noise: A Review and Bibliography", "Robotics Bibliography 1970–1981", "Machine Intelligence – An International Bibliography with Abstracts on Sensors in Automated Manufacturing", and "Robotics: An International Bibliography with Abstracts".

Contents

1. GENERAL LITERATURE
   1.1 Books   1
   1.2 Proceedings   2
   1.3 Bibliographies   3
   1.4 State-of-the-art reports/reviews   5

2. LASER SYSTEMS
   2.1 Types and optical equipment   15
   2.2 Lasers and robots   24
   2.3 Lasers and advanced manufacturing systems   28

3. LASER BEAM PHYSICS
   3.1 General   33
   3.2 Laser optics   36
   3.3 Laser thermodynamics and energy   39
   3.4 Plasma phenomena and plumes   45
   3.5 Laser system modelling   49

4. LASER PROCESSES
   4.1 Laser welding
       4.1.1 General   53
       4.1.2 Laser welding of metals
             4.1.2.1 General   59
             4.1.2.2 Laser welding of ferrous metals   63
             4.1.2.3 Laser welding of non-ferrous metals   69
       4.1.3 Laser welding of dissimilar metals   72
   4.2 Laser cutting
       4.2.1 General   75
       4.2.2 Laser cutting of metals   78
       4.2.3 Laser cutting of non-metals   85
       4.2.4 Laser drilling
             4.2.4.1 General   88
             4.2.4.2 Laser drilling of metals   88
             4.2.4.3 Laser drilling of non-metals   90
       4.2.5 Laser machining
             4.2.5.1 General   90
             4.2.5.2 Laser machining of metals   92
             4.2.5.3 Laser machining of non-metals   94

4.3 Laser surface modification
   4.3.1 General   97
   4.3.2 Coatings and films   98
   4.3.3 Cladding and alloying   99
   4.3.4 Heat treatment and surface hardening   104
   4.3.5 Etching, marking and scribing   111
4.4 Metallurgical analysis of laser processes
   4.4.1 General   114
   4.4.2 Metallurgy of laser welding and processing of
          ferrous metals and alloys   114
   4.4.3 Metallurgy of laser welding and processing of
          non-ferrous metals and alloys   121
4.5 Laser processing safety measures   127

5. INDUSTRIAL APPLICATIONS OF LASERS
5.1 Industrial and aerospace applications   135
5.2 Nuclear and power industries   141
5.3 Electronics industries   146
5.4 Testing and inspection   155

AUTHOR INDEX   159

Introduction

Research workers in the early 1960s found that a ruby laser could easily melt and vaporise small amounts of metal. Subsequently many investigations have been performed to determine the effects of high-power laser radiation on absorbing surfaces. By the late 1960s lasers had become practical production tools, and in the early 1970s advances in the power of CO_2 lasers led to deep penetration laser welding. This greatly increased the range of metal thickness amenable to laser processing.

By the 1980s the four primary laser applications existing in production within a variety of industries are deep-penetration laser welding, high precision laser cutting, surface heat treatment by martensitic phase transformation hardening, and surface alloying.

Applications in the traditional production engineering industries are rapidly being overtaken by developments in the use of lasers in the electronics industries where the high precision and quality of laser welding and cutting is seen at its most beneficial. Areas such as laser assisted crystal regrowth and annealling of iron implantation damage, point towards new methods of generating semiconductor circuitry.

This comprehensive bibliography covers all the major applications and includes substantial material on the physics and metallurgy of both CO_2 and neodymium–yttrium-aluminium-garnet (Nd:YAG) laser processing. Processing of both metals and non-metals is covered and a substantial section is concerned with laser processing safety.

Care has been taken to select only those references which are available from publishers, and national and industrial libraries; and it is hoped that a balance has been achieved in the coverage of US, European and Japanese information on this rapidly developing technology.

Alan Gomersall
September 1986

1

GENERAL LITERATURE

1.1 BOOKS

1.1 (1) Laser machining and welding

Kokora, A., Rykalin, N. and Uglov, A. Pergamon Press, New York, 1979, 310pp.

The basic physical effects of laser radiation on opaque mediums – primarily metals – are first described. CO_2 laser heat treatment and welding are covered and various thermophysical properties related to laser effects are considered. Data on drilling of materials are given and the motion of the liquid phase during formation of a hole is analysed, together with factors which affect drilling and precision and reproducibility of laser machining. Changes in metal structure and properties in the zones affected by laser radiation are covered, including the effects of the Q–switched laser radiation on metals. Appendices include improvements in laser drilling techniques, effect of atmosphere on laser machining, surface hardening of machine parts, conical lens for drilling of large-diameter holes, and cutting of non-metallic materials with CO_2 laser beams. (213 refs.)

1.1 (2) Lasers in electron beam processing of materials

White, C. W. and Peerey, R. S. (Eds.). Academic Press, 1980, 769pp.

1.1 (3) Source book on applications of the laser in metalworking

Metzbower, E. A. American Society for Metals, Metals Park, OH, USA, 1981, 387pp.

1.1 (4) Laser materials processing

Bass, M. (Ed.) (University of Southern California, USA). *Materials Processing, Theory and Practice*, Vol. 3. North-Holland, Amsterdam, 1983, 490pp.

The book is devoted to discussions of some of the most important materials processing activities in current use or development. The articles describe the general mechanisms and principles responsible for the processes. Topics considered include: laser drilling, milling and machining, semiconductor manufacture, laser welding, and steel heat treatment.

1.1 (5) Lasers in metalworking – A summary and forecast

Sanderson, R. J. Tech Tran Corp., Naperville, IL, USA, 1983, 165pp.

1.1 (6) Laser welding, cutting and surface treatment

Crafer, R. C. Welding Institute, Cambridge, UK, 1984, 58pp.

Discussions include laser equipment trends, laser beam parameters, analysis of machines tool systems suitable for laser profiling, laser surface treatment of metals, and the use of lasers in the aerospace, automotive, and nuclear industries. The majority of industrial applications are carried out with Nd: YAG or CO_2 lasers. (102 refs.)

1.1 (7) Laser and plasma technology

Lee, S. et al. World Scientific Publishing, Singapore, 1985, 696pp.

1.2 PROCEEDINGS

1.2 (1) Applications of lasers in materials processing

Metzbower, E. A. (Ed.), 18-20 April 1979, Washington, DC. American Society for Metals, Metals Park, OH, USA, 1979, 330pp.

Contents include: Fundamentals of interactions; Metal reflectance intense laser irradiation; Structure of the evaporation front from Boltzmann equation; Pulsed YAG laser welding of ODS alloys; Laser welding of exhaust gas oxygen sensor; Element analysis of laser welding induced thermal shock; Mechanical properties, fracture toughnesses and laser welds of high strength alloys; Laser materials fabrication; Shaping continuous wave carbon dioxide laser; Measurement technique for controlling dimensions of radially forged cylinders; Laser beam detection of ultrasonic and acoustic emission signals for non-destructive testing of materials; Melting and solidification – Moving heat flux; Microstructural transformations by the laserglaze process in Zircaloy-4 sheet; Solidified, laser surface alloyed, low-carbon steels; Processing and properties of laser surface alloys; Production laser cutting; Interaction of laser-induced stress waves; laser shock processing of aluminium alloys.

1.2 (2) Advanced processing methods for titanium

Hasson, D. F. (Ed.) (US Naval Academy), 13-15 October 1981, Louisville, KT. Metallurgical Society of AIME, Warrendale, PA, USA, 1982, 315pp.

Contains 18 papers arranged in the following categories: bulk forming; forming; casting and welding; and machining. Laser welding and the difficulties of machining titanium alloys are among the topics discussed.

1.2 (3) ICALEO '82 – International congress of applications of lasers and electro-optics

LIA, 20-23 September 1982, Boston, MA. Laser Institute of America, Toledo, OH, USA, 1982, 524pp.

Proceedings include 96 papers presented in five volumes dealing with: materials processing; medicine and biology; inspection, measurement and control; lasers and electro-optics; and optical communications. Topics considered include: welding and soldering laser cutting and drilling, and safety.

1.2 (4) Industrial applications of laser technology

Fagan, W. F. (Ed.) AOL-Dr. Schuster GmbH, Austria). 19-22 April 1983, Geneva. SPIE, Bellingham, WA, USA, 1983, 422pp.

Contains 61 papers containing information on holographic interferometry; laser measurement techniques; laser material processing. The topics discussed include: optical components; material testing; contouring of cutting tools; surface measurements; position measurements; laser Doppler velocimetry; laser cutting and laser annealing.

1.2 (5) Lasers in materials processing

Metzbower, E. A. (Ed.) US Naval Research Lab., USA). 24-26 January 1983, Los Angeles. American Society for Metals, Metals Park, OH, USA, 1983, 272pp.

Contains 26 papers on: pulsed laser annealing of semiconductors; laser marking techniques; metal cutting with laser beams; hardfacing with laser beams; laser surface treatment; laser surface alloying; laser melting; laser processing of plasma-sprayed coatings; and laser welding and soldering.

1.2 (6) Application of lasers in mechanical-engineering technology

11-13 October 1982, Zvenigorod, USSR. *Bull. Acad. Sci. USSR Phys. Ser.*, Vol. 47, No. 8, 1983, 93pp.

Contains 14 papers on laser applications in mechanical engineering, including surface modification, physical processes, laser cutting technology, alloying surface layers, technological lasers, optical transmission elements, metallic optics, laser light telemetry, and radiation sources.

1.2 (7) Applications of high power lasers

22-23 January 1985, Los Angeles, SPIE, Bellingham, WA, USA, 1985.

The following topics were dealt with: laser beam cutting, machining and welding, heat treatment, laser cladding techniques, and laser beam machining in the microelectronics industry. Individual papers are exhibited in relevant sections in this bibliography.

1.3 BIBLIOGRAPHIES

1.3 (1) Laser welding

Reed, W. E. (Ed.) (NTIS, USA). Citations from the Engineering Index Data Base (1970-1977). NTIS/PS-77/0816/7EES. National Technical Information Service, Springfield, VA, USA, 1977, 214pp.

Citations deal with the development of techniques and procedures for laser welding and with the characteristics of the resulting welds. Process control, including automation and applications of welding of stainless steels, dissimilar metals, titanium, ceramics, and plastics.

1.3 (2) Laser welding

Reed, W. E. (Ed.) (NTIS, USA). Citations from the Engineering Index Data Base (1970-1980). NTIS/PB80-815202. National Technical Information Service, Springfield, VA, USA, 1980, 306pp.

Citations deal with the development of techniques and procedures for laser welding and with the metallurgical characteristics of the resulting welds. Process control, automation, and applications including welding of stainless steels, dissimilar metals, titanium, ceramics, and plastics.

1.3 (3) Neodymium glass lasers

Young, C. G. (Ed.) (New England Research Application Center, USA). Citations from the Engineering Index Data Base (1970-1980). NTIS/PB80-809213. National Technical Information Service, Springfield, VA, USA, 1980, 161pp.

Includes 154 references to the design, construction, and uses of neodymium glass lasers including drilling, cutting, and welding of materials.

1.3 (4) Laser welding

Citations from the Engineering Index Data Base (1979-1982). NTIS/PB83-806117. National Technical Information Service, Springfield, VA, USA, 1983, 348pp.

341 references are given which deal with the development of techniques and procedures for laser welding and with the characteristics of the resulting welds. Process control, automation, and applications are also listed. Materials welded include stainless steels, dissimilar metals, titanium, ceramics, and plastics.

1.3 (5) Laser beam welding: nickel, titanium, and their alloys

Citations from the Metals Abstracts Data Base (1966-1985). NTIS/PB85-858348/XAD. National Technical Information Service, Springfield, VA, USA, 1985, 93pp.

Contains 172 references concerning laser beam welding technology relative to titanium and nickel and their alloys. Controlling variables in the laser beam welding process for efficiency of use for industrial applications, equipment expenses, and comparisons and contrasts of laser beam welds to other weld processes are among the topics discussed.

1.3 (6) Laser beam welding: steels

Citations from the Metals Abstracts Data Base (1966-1984). NTIS/PB85-867935/XAD. National Technical Information Service, Springfield, VA, USA, 1985, 144pp.

Contains 279 citations concerning research and development of laser beam welding of steel alloys. Topics include beam control techniques, weld comparisons with conventional techniques, industrial applications and compatibility studies, and weld performance evaluations. Applications to a wide range of thicknesses of various steel alloys are discussed.

1.3 (7) Laser beam welding: steels

Citations from the Metals Abstracts Data Base (1984-1985). NTIS/PB85-867943/XAD. National Technical Information Service, Springfield, VA, USA, 1985, 58pp.

Contains 74 references to research and development of laser beam welding of steel alloys. Topics include heat transfer studies, beam focusing techniques, and weld penetration investigations. Emphasis is on CO_2 laser devices.

1.3 (8) Laser cutting and machining

Citations from the US Patent Data Base (1970-1975). NTIS PB85-862530/XAD. National Technical Information Service, Springfield VA, USA, 1985, 52pp.

Contains 58 references to selected patents concerning methods and equipment description of laser machining in industrial manufacturing. Laser machining device and process design techniques, workpiece protection, and slag removal during laser machining operations are included.

1.3 (9) Laser cutting and machining: metals

Citations from the Engineering Index Data Base (1970-1985). NTIS/PB85-861805/XAD. National Technical Information Service, Springfield, VA, USA, 1985, 148pp.

Contains 243 citations concerning laser applications in metalworking. Topics include development and technological reviews, theoretical studies, and product quality evaluations. The utilisation of laser devices for welding, cutting, drilling, and scribing a variety of metals is discussed. Heat transfer and the effects on machined materials, and economic aspects of laser metalworking are also considered.

Contains 121 references to technological innovations in laser beam welding of aluminium base alloys. Weld seam fusion penetration, control systems for beam power, focused power density and travelling velocity, equipment operation, maintenance, and expense, and applications for laser beam welding of aluminium alloys are among the topics discussed. Metallurgical papers are also referenced.

1.3 (10) Laser cutting and machining: non-metals

Citations from the Engineering Index Data Base (1970-1985). NTIS/PB85-862357/XAD. National Technical Information Service, Springfield, VA, USA, 1985, 71pp.

Contains 107 references to the utilisation of laser techniques and equipment in the machining and cutting of a variety of non-metallic materials, including wood, plastics, ceramics, glass, paper, and diamonds. Descriptions of specific drilling, welding, and shaping operations are included. The use of carbon dioxide and YAG lasers is also included.

1.3 (11) Laser maching of thin metal films and refractory materials

Citations from the Metals Abstracts Data Base (1966-1985). NTIS/PB85-857894/XAD. National Technical Information Service, Springfield, VA, USA, 1985, 81pp.

Contains 156 references to the use of lasers as a machining tool. The emphasis is on precision cutting and drilling of thin metal films while minimising distortion and heating effects. The laser machining of refractories and hard materials, such as silicon carbide, is included.

1.3 (12) Laser metal cutting

Citations from the Metals Abstracts Data Base (1966-1985). NTIS/PB86-853074/XAD. National Technical Information Service, Springfield, VA, USA, 1985, 165pp.

This bibliography contains 353 references to laser metal cutting. High precision and CO_2 laser cutting methods are considered. Three-dimensional cutting process and the cutting of thin metal sheets without edge deformation are included.

1.3 (13) Laser welding: aluminium

Citations from the Metals Abstracts Data Base (1966-1985). NTIS/PB86-850955/XAD. National Technical Information Service, Springfield, VA, USA, 1985, 69pp.

1.3 (14) Laser welding

Citations from the US Patent Data Base (1970-1985). NTIS/PB85-867471/XAD. National Technical Information Service, Springfield, VA, USA, 1985, 74pp.

Contains 88 references to selected patents concerning development and innovations in laser welding techniques, processes, and equipment. Laser beam control, clamping devices, welding jigs, weld seam tracking, production laser welding systems, and work station enclosures and safety devices are among the topics discussed.

1.3 (15) Laser welding

Citations from the Engineering Index Data Base (1982-1985). NTIS/PB86-850609/XAD. National Technical Information Service, Springfield, VA, USA, 1985, 79pp.

Bibliography contains 113 references to laser welding techniques, and evaluations of the welded joints in alloys and plastics that can be laser welded. Some references cover automation of laser welding, and comparisons of laser welding with other welding techniques.

1.3 (16) Neodymiun YAG lasers

Citations from the INSPEC Data Base (1975-1983). NTIS/PB85-871960/XAD. National Technical Information Service, Springfield, VA, USA, 1985, 363pp.

Contains 313 references concerning the properties and applications of neodymium-yttrium-aluminium-garnet lasers. Applications include welding of parts such as printed circuits, medical applications, telecommunication systems, and optical pumping for high powered lasers.

1.3 (17) Neodymium YAG lasers

Citations from the INSPEC Data Base (1983-1985). NTIS/PB85-871978/XAD. National Technical Information Service, Springfield, VA, USA, 1985, 134pp.

Contains 145 references on similar topics to the previous reference.

1.4 STATE-OF-THE-ART REVIEWS/REPORTS

1.4 (1) Laser machining and fabrication – a review

Scott, B. F. (University of Birmingham, UK). In, *Proc. 17th Machine Tool Design and Research Conf.*, Vol. 2 – *Machine Tool Vibration and Noise*, 20-24 September 1976, Birmingham, UK, pp. 335-339. Sponsored by University of Birmingham and University of Manchester Institute of Science and Technology.

The machining and welding characteristics of solid-state and nitrogen-carbon dioxide lasers are discussed on the basis of their ability to create an appropriate level of the absorbed power intensity in the workpiece. Methods for the prediction of performance and for the establishment of machining standards are considered on the same basis. Limitations and design problems of lasers are identified. (15 refs.)

1.4 (2) Pulsed-laser metalworking

Bolin, S. R. (Raytheon Co., USA). *American Machinist*, 20(10): 123-126, 1976.

Discusses the applications of pulsed lasers in welding and drilling, offering some practical examples. Three typical applications of pulsed-laser welding are detailed: wire-to-terminal welding, seam welding, and mixed spot and seam welding. Laser drilling of diamond wire-drawing dies and the cooling holes in jet turbine components is a typical application briefly described.

1.4 (3) Welding and cutting with laser beams – considerations concerning the state-of-the-art

Beck, R. (Battelle Institute, West Germany). *Schweissen und Schneiden*, 29(5): 170-172, 1977. (In German)

Discusses the laser types suited for welding and cutting of metals and the power levels attainable and describes welding of workpieces less than 1mm. thick and of those 1 – 30mm. thick. Gas jet supported laser cutting is briefly discussed, and cutting speeds attained with aluminium, stainless steel, and titanium sheet are tabulated.

1.4 (4) Applications of pulsed solid-state lasers

Naugler, T. W. (Raytheon Co, USA). In, *Proc. Tech. Programme Electro-Optic Laser Conf. Exposition*, 19-21 September 1978, Boston, MA, pp. 102-107. Industrial and Science Conference Management Inc, Chicago, 1978.

Discusses specific applications in the areas of laser cutting, hole drilling, heat treating, reflow soldering, brazing, and deoxidising; all of which can justify the introduction of solid-state pulsed laser system processing. Laser parameters, process speeds, and parts handling considerations are included.

1.4 (5) Material processing with the laser: The principles and present position of the technique

Keck, R. and Wilhelm, H. G. *Stahl, Eisen*, 98(13): 665-669,1978. (In German)

Reviews the use and advantages of the CO_2 laser for machining, welding, cutting, and surface hardening of metals.

1.4 (6) What designers should know about laser welding and cutting

Anderson, D. G. (Union Carbide Corp, USA). *Mechanical Engineering*, 100(6): 44-49, 1978.

The high-power CO_2 laser has been applied successfully to a variety of welding and cutting applications. Emphasises advantages such as high speed, reliable operation, and quality of finished product. Discusses the principles of laser processing, the advantages and limitations of laser welding and cutting, a number of industrial welding and cutting applications, and describes two examples of welding/cutting systems by laser.

1.4 (7) The laser: a cutting and welding tool

Sharp, C. M. *Mach. Mod. (Paris)*, 840: 89-92, 1979. (In French)

The use and performance of the laser as a cutting and welding tool and production laser are described. Operating at a power density of $105\text{-}106W/cm^2$, the cutting process is accompanied by gas blowing to ensure a narrow fissure. An inert gas is used to avoid oxidation or a reactive gas such as oxygen in cutting steel. Quality decreases with increased thickness of the workpiece. The advantages of laser cutting include the lack of wear on the cutting tool, high cutting speed, and minimal heating of the workpiece and material loss. The hardness of the material does not affect the rate of cutting and welding of the workpiece can follow immediately without any further preparation. Using a lens with a long focal length, several layers of material may be cut simultaneously.

1.4 (8) Cost reduction through manufacturing technology

Slack, R. B. (Pratt & Whitney, USA). In, *16th Jet Propulsion Conf.*, 30 June- 2 July 1980, Hartford, CT, USA, Paper 80-1251, 8pp. AIAA, New York, 1980.

A brief discussion of the cost reduction impacts and potentials of new technologies is presented. Includes electron beam hole drilling and laser welding. In each case, a technical review and specific cost reduction experience are reviewed. Some of these new technologies have resulted in labour cost reductions of up to 90%.

1.4 (9) Fine drilling and cutting: Solid pulsed laser

Weedon, T. M. W. and Wright, J. K. *Machine Moderne*, 848 (September): 191-194, 1980. (In French)

Brief examples are given of the YAG laser's application in thermal treatment, welding, drilling, welding to a depth greater than 1mm, and cutting.

1.4 (10) Laser processing of materials

Kear, B. H. In, *Materials Aspects of World Energy Needs Conf.*, 26-30 March 1979, Reston, VA, USA, pp. 364-366. National Academy of Sciences, Washington, DC, 1980.

Lasers, in addition to conventional applications are now being considered for laser glazing. This process involves surface localised melting followed by rapid solidification and subsequent solid state cooling. Enhanced corrosion, erosion and wear properties may be achieved by melting in pre-alloyed material. Layer glazing is a related development in which a bulk rapidly solidified structure is formed by the sequential build up of one laser glazed layer upon another using a continuous supply of pre-alloyed feed material. Laser shock hardening which uses very high power density pulsed lasers to effect surface hardening from the blast wave produced from the rapid expansion of the vapourised material is also considered.

1.4 (11) Laser welding, cutting and drilling

Bolin, S. (Raytheon Co, USA). *Assembly Engineering*, 23(5): 30-34, 1980.

Deals with pulsed solid state (primarily YAG) lasers and their material processing applications. It covers characteristics of the technique and its applications and materials selection factors.

1.4 (12) Progress in researches on laser material processing in Japan

Kobayashi, A. *Bull. Jpn. Soc. Precision Engineers*, 14(4): 201-206, 1980.

During the period 1963-79, 142 papers on the processing of metals and other materials by ruby, CO_2, and YAG, lasers were presented at Annual Meetings of the Japan Society of Precision Engineers. The reports are reviewed according to the material treated, the treatment concerned, the laser technique employed, and the industrial, laboratory, are academic sources of the research. The papers were concerned with drilling, welding, and trimming or cutting. Similar reviews and classifications are given for another 173 papers published in Japan from 1970-79, and industrial applications of the lasers are discussed in the light of the overall data.

1.4 (13) Laser processing of materials

Breinan, E. M., Kear, B. H. and Thompson, E. R. In, *Advances in Metal Processing, 25th Sagamore Army Materials Research Conf.*, 17-21 July 1978, Lake George, NY, USA, pp. 45-78. Plenum Press, New York, 1981.

Presents an overview of developments in laser processing of materials, including transformation hardening (gray iron) and surface alloyings welding of X80 steel, cutting of Al and Ti, drilling, rapid solidification processing, pulse annealing, shock hardening, laser-assisted machining and laser-controlled surface reactions. Several current industrial applications are described. (34 refs.)

1.4 (14) Application of a high-power laser to demilitarisation problems

Lyman, O. R. (Army Armament Research and Development Command, USA). Report No. ARBRL-TR-02360, 34pp., 1981.

Discusses attempts to experimentally cut rocket motors, puncture agent cavities, and perform other operations associated with demilitarisation of munitions with a high-power CO_2 laser, and requirements are outlined and problems associated with laser operations contiguous with hazardous materials are highlighted.

1.4 (15) Application pointers from laser development for industry

Sharp, M. *Stainless Steel Industries*, 9(50): 13-15, 1981.

Discusses some of the advances made in the last seven years in the development of laser equipment for industry. Mention is made of the advantages of laser machining, the benefits of piped power beams, cost savings, cutting rates applicable to stainless steel and welding.

1.4 (16) Laser beam processing – A manufacturer's viewpoint

Osterink, L. and Peng, Y. C. J. In, *Wear and Fracture Prevention, Conf. Proc.*, 21-22 May 1980, Peoria, IL, USA, pp. 135-143. American Society for Metals, Metals Park, OH, USA, 1981.

The four primary laser applications existing in production within a variety of industries are deep-penetration laser welding, high precision laser cutting, surface heat treating by martensitic phase transformation hardening and surface alloying. Each area is discussed in detail and the advantages and various parameters to achieve maximum productivity and quality are described. Beam configuration, integration and manipulation are included. Production examples of laser welding, cutting, surface hardening and surface alloying are examined.

1.4 (17) Laser cutting for welded joints

Hill, M., Johnson, R. and Megaw, J. H. C. In, *The Joining of Metals: Practice and Performance, Conf. Proc.*, 10-12 April 1981, Coventry, UK, Vol. 1, pp. 1-8. Institution of Metallurgists, London, 1981.

Laser welding can operate in a keyholing mode and produce welds with almost parallel sides. The weld preparation for this process can be thus a simple square butt design, and this can be achieved by laser cutting. The cuts are almost parallel-sided and although the cutting gas flow creates surface rippling along the cut edges this can be quite shallow and so permit good weld fit-up. An investigation into optimising the laser cutting process for rewelding has been started, and the effects of altering several of the process variables are discussed. From the limited re-welding trials to date the use of high power lasers for cutting and rewelding applications appears promising.

1.4 (18) The new wave machine tool

Bushor, W. E. *Metalwork*, March-April 1981: 5-8.

Lasers are increasingly used for such operations as drilling, cutting and welding. Lasers used include ruby, Nd:YAG, Nd:glass and CO_2; each laser system has characteristics that make it better suited for one or more of the above operations. Basic principles of laser machining are reviewed and examples of individual applications are discussed. Advantages of laser machining include precise metal removal, narrow kerf width, ability to process hard-to-machine materials, high efficiency and cost

savings. Lasers have also been used for resistor trimming, marking, engraving, sealing and heat treating.

1.4 (19) Fast axial flow carbon dioxide laser materials processing

Oakley, P. J. In, *ICALEO '82, Vol. 31, Materials Processing Conf. Proc.*, 20-23 September 1982, Boston, MA, pp. 120-128. Laser Institute of America, Toledo, OH, USA, 1982.

Current research into laser materials processing using these machines can be divided into: laser surfacing, high-speed laser welding of sheet materials (low-alloy and carbon steels) and the mechanical properties of laser welds. These research programmes are described and significant results are highlighted. A plasma control technique for welding at relatively slow speeds, developed at the Welding Institute, is described. (9 refs.)

1.4 (20) The laser: An industrial tool

Steen, W. M. *Met. Soc. World*, 1(7/8): 12-13, 1982.

Developments in the use of lasers for materials processing are reviewed. The use of polarised laser beams for metal cutting has increased the speed of cutting and quality of the cut edge. Numerically controlled laser cutting can now rapidly produce complex shapes. Improved control of laser welding techniques have been achieved recently and the coupling of TIG and laser welding/cutting now provides a cheap and rapid welding/cutting process without significant loss of quality. The speed of operation enables a fine-grained weld with a low HAZ to be formed.

1.4 (21) The laser beam as a thermal tool in production technology

Herbrich, H. and Weskott, D. *Ind. Anz.*, 104(69): 25-29, 1982. (In German)

Discusses the industrial application of laser beams in conventional metalworking processes such as sawing, punching and nibbling. Presented and described are: CO_2 gas laser function principles, gas unloading system, laser beam cutting, mobile use of CO_2 lasers on control machines, applications for laser cutting and the outlook for welding and surface treatment.

1.4 (22) Laser treatment in mechanical engineering

Nikolaev, G. A. and Grigor'yants, A. G. (Moscow Technical College, USSR). *Bulletin Acad. Sci.,*

47(8): 1-9, 1983. Presented at All-Union Conf. on the Applications of Lasers in Mechanical Engineering Technology, 11-13 October 1982, Zvenigorod, USSR.

Discusses the use of laser beams in processes of welding, cutting, melting, alloying, and heat treatment of surfaces. Basic features of laser welding, as well as fusional ability of laser radiation and the results of experimental and theoretical investigations are considered. Also outlines the use of lasers for the heat treatment and chemicothermal treatment of steel.

1.4 (23) Lasers in manufacturing

Weld. Rev., 1(3): 37-38, 1982.

Lasers give high intensity, coherent and monochromatic light beams, are clean, with no combustion of ion bombardment problems, and the heat is very localised, reducing thermal damage. Laser surface treatment can offer transformation hardening, rapid surface quenching, cladding, and surface alloying. Laser cutting and welding principles and applications in industry are described.

1.4 (24) Lasers in production

Sharp, C. M. and Steen, W. M. *Eng. Mater. Des.*, 26(2): 32-36, 1982.

High-powered CO_2 lasers produce multikilowatts of optical energy with a commercially acceptable level of reliability in service – approx 90% for the controlled laser 2kW CO_2 laser – and a high level of power stability over long periods. By optically focusing the laser beams onto the workpiece, the photon energy is absorbed and converted into heat for submelting, melting or vapourisation. The laser is also able to cut through cloth, rubber, wood, steel, Al and brass etc, by evaporating a hole and blowing the molten material away with a gas jet. Laser welding is similar to cutting except the vapour keyhole is allowed to solidify under an inert gas shield to produce a more homogeneous and finer structure in the fusion zone with high joint efficiency and deeper penetration.

1.4 (25) Material processing – an overview

Ready, J. F. (Honeywell Corporate Technology Center, USA). *IEEE Proc.*, 70(6): 533-544, 1982.

Laser researchers in the early 1960s found that a ruby laser could easily melt and vapourise small amounts of metal. Many investigations have since been performed to determine the effects of high-power laser radiation on absorbing surfaces. By the late 1960s lasers had become practical production tools. In the early 1970s, advances in the power of CO_2 lasers led to deep penetration laser welding. This greatly increased the range of metal thickness amenable to laser processing. At present, lasers are used in a practical way for many applications in material processing. For some applications, like trimming of resistors and drilling of holes in ceramics, laser processing is the leading method. For other applications, like heat treating, welding, and cutting, laser processing is an economically competitive alternative to conventional techniques. In addition there are new research possibilities, especially for processing of semiconductors. Areas such as laser-assisted crystal regrowth and annealing of ion-implantation damage point toward new methods of generating semiconductor circuitry. The physical phenomena which underlie laser processing applications are reviewed.

1.4 (26) Use of gaseous halogen compounds in fusion welding. (A survey of literature)

Bad'yanov, B. N. et al. *Svar. Proizvod*, 4: 16-17, 1982. (In Russian)

In welding, inert gases (Ar, helium) or active gases (CO_2, nitrogen) or their mixtures are used for weld protection. Halogen compounds increase the depth of welding fusion in steels and Ti alloys. The halide-containing protective gases increase the depth of metal fusion in laser welding, decrease the impact strength of the welds and deteriorate the hygienic conditions in the welding area.

1.4 (27) Application of solid state lasers in manufacturing industry

Weedon, T. M. (UK Lasers Ltd, UK). In, *Proc. 1st Int. Conf. on Lasers in Manufacturing*, 1-3 November 1983, Brighton, UK, (Ed. M. F. Kimmitt), pp. 1-12. IFS (Publications) Ltd, Bedford, UK, 1983.

YAG lasers are established production tools in a wide range of industries. They are used for drilling, welding, cutting and heat treatment. They compete on economic as well as technical grounds with the traditional processes for these activities. This paper concentrates on the use of pulsed YAG lasers for welding, drilling and cutting. Welding is applicable to joints from 0.010 to 3mm. thick. Drilling and cutting are applicable to material from 0.001 to 10mm. thick.

1.4 (28) Considerations for lasers in manufacturing

Charschan, S. S., Webb, R. and Bass, M. (Western Electric Co, USA). In, *Laser Materials Processing*, pp. 439-473. North-Holland, Amsterdam, Netherlands, 1983.

Considers technical feasibility and economic justification. Assuming technical feasibility to be established, attempts to place in perspective the other considerations. Guidelines are provided for economic justification; and given sample specifications for system performance and evaluation, the preparation required for system installation, operation, and maintenance is reviewed. Recommendations are made relative to spare parts and servicing; and the skills and training requirements of associated personnel, are explained. (19 refs.)

1.4 (29) Evaluating a CO_2 industrial laser system

Laos, O. V. (Laser Scientific Services Ltd, UK). In, *Proc. 1st Int. Conf. on Lasers in Manufacturing*, 1-3 November 1983, Brighton, UK, (Ed. M. F. Kimmitt), pp. 21-30. IFS (Publications) Ltd, Bedford, UK, 1983.

This paper on "How to and why select a CO_2 industrial laser system" avoids complex laser theory. It is for the appraisal of a potential laser user as an introduction into the world of lasers.

1.4 (30) High PRF nitrogen-carbon dioxide laser for continuous manufacturing processes in metals

Chatwin, C. R. and Scott, B. F. (University of Glasgow, UK). In, *Proc. 1st Int. Conf. on Lasers in Manufacturing*, 1-3 November 1983, Brighton, UK, (Ed. M. F. Kimmitt), pp. 263-272. IFS (Publications) Ltd, Bedford, UK, 1983.

Some fundamental limitations and design problems of gas and solid state lasers are identified. The possibility of achieving continuous manufacturing processes in metals with a pulsed gas laser is considered because this mode of operation removes intensity limitations inherent in CW operation.

1.4 (31) Laser based manufacturing – CO_2 systems

Rourke, M. (Control Laser Ltd, UK). In, *Proc. 1st Int. Conf. on Lasers in Manufacturing*, 1-3 November 1983, Bighton, UK, (Ed. M. F. Kimmitt), pp. 231-241. IFS (Publications) Ltd, Bedford, UK, 1983.

Gives an insight into the areas in which lasers are now established and the selection of machine tool packages that can be created around lasers, developing up to 4kW of power using mainly computer numerical control (CNC).

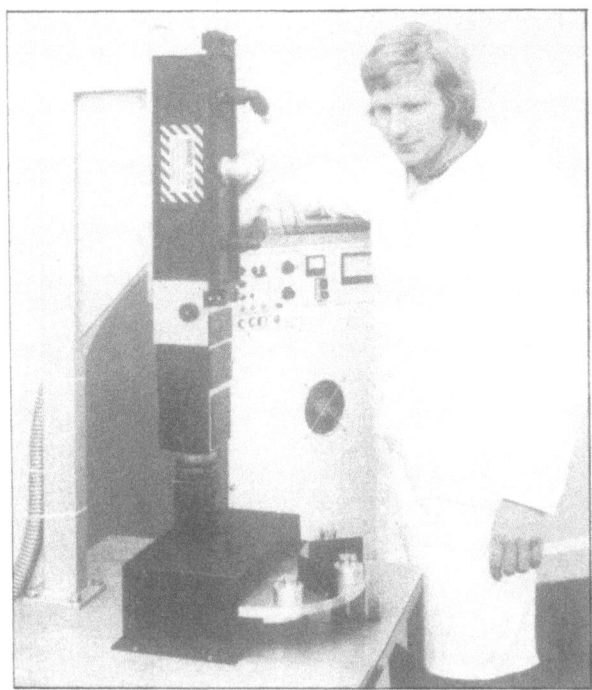

'CLAM' compact laser machining system with a fabricated stand and a simple indexing table (See 1.4 (27))

1.4 (32) State of laser-cutting technology and prospects for its development

Tikhomirov, A. V. (All-Union Scientific Research Institute of Autogenous Mechanical Engineering, USSR). *Bulletin Acad. Sci.*, 47(8): 23-28, 1983. Presented at All-Union Conf. on the Applications of Lasers in Mechanical Engineering Technology, 11-13 October 1982, Zvenigorod, USSR.

An analysis is given of the 1983 state of theoretical developments, technological processes, and machines for laser welding, and of prospects for their development. Radiation parameters of specialised lasers are formulated.

1.4 (33) Survey of high-power CO_2 industrial laser applications and latest laser developments

Eberhardt, G. (Spectra-Physics GmbH, W. Germany). In, *Proc. 1st Int. Conf. on Lasers in Manufacturing*, 1-3 November 1983, Brighton, UK, (Ed. M. F. Kimmitt), pp. 13-21. IFS (Publications) Ltd, 1983.

Focuses on typical applications in cutting, welding and heat treating for low volume, as well as mass production. Manual load/unload as well as automated systems will be discussed. Also describes recent CO_2 laser developments incorporating user requirements of higher stability, increased reliability, simpler operation and

Two-station welder with two focusing heads for each station (See 1.4 (33))

improved ruggedness. These improvements allow the use of unskilled people to operate the laser; further the user's maintenance crew is able to perform routine maintenance and service.

1.4 (34) Trends in the development of the application of CO_2 lasers in materials technology

Mordike, B. L. (Technische University Clausthal, W. Germany). *Z Werkstofftech.*, 14(7): 221-228, 1983.

Demonstrates the applications of CO_2 lasers in materials processing. Examples in laser cutting, laser welding, surface heat treatment and surface alloying are discussed.

1.4 (35) Application of CO_2 lasers in mechanical engineering technology in the USSR

Abil'siitov, G. A. and Velikhov, E. P. (Academy of Sciences of the USSR). *Opt. Laser Technology*, 16(1): 30-36, 1984.

The various types of lasers in use are investigated and the main properties of laser welding and cutting are studied and compared with other forms of these processes. Methods of surface treatment and heat hardening are also investigated.

1.4 (36) Industrial applications of YAG lasers

Decaux, J.-M. (Optilas, France). In, *Opto '84, Proc. 4th Optoelectronic Meeting*, 15-17 May 1984, Paris, France, pp. 82-84. ESI Publications, Paris, France, 1984. (In French)

Used in the Q-switched mode CW YAG lasers are mainly devoted to surface evaporation techniques such as: cutting of thin materials, resistor trimming, marking and surface treatment. Pulsed YAG lasers have a higher penetration and are used in welding, drilling/cutting and marking.

1.4 (37) Laser metalworking

Saunders, R. J. *Metal Progress*, 126(2): 45, 47-48, 50-51, 1984.

Presents advantages, disadvantages and some considerations in cutting with lasers. The meaning of mode burn, polarity and their implications are explained, and multimode and single-mode lasers are compared. The roles of and the relationship between metal thickness and cutting speed for low-carbon and stainless steels, Al-, Ti- and Ni-base alloys, are presented. Also discusses some important aspects of laser welding.

1.4 (38) Lasers in the nuclear industry

Johnson, R. (Welding Institute, Cambridge, UK). In, *Laser Welding, Cutting and Surface Treatment*, pp. 43-47. Welding Institute, Cambridge, UK, 1984).

The United Kingdom Atomic Energy Authority is currently using lasers with power capabilities of less than 1mW to more than 15kW continuous wave output, together with ultra-high energy single-pulse lasers, in the plasma research associated with fusion power. The lower power lasers are used for inspection, alignment, and diagnostic purposes. 100W-15kW lasers are in operation in production plants or in research laboratories to assess the feasibility of further production uses. Systems in use and applications are described. (7 refs.)

1.4 (39) Look to lasers for a practical way to do the tough fabricating jobs

Smoluk, G. R. *Mod. Plastics*, 61(5): 61-63, 1984.

The discussion covers uses of lasers; new design flexibility, equipment advances; setting-up lasers; and economics of laser materials processing.

1.4 (40) Designing parts for laser processing

Gregson, V. G. (Laser Center, Westinghouse Marine Division, USA). In, *Proc. SPIE Conf. on Applications of High-Power Lasers*, Vol. 527, 22-23 January 1985, Los Angeles, CA, pp. 73-79. International Society of Optical Engineers, USA, 1985.

The movement of a laser-processed part on to the manufacturing floor is a problem of parts design. The steps which must be completed before a part can be described as designed for laser processing are described.

1.4 (41) The development of laser technology and its importance for material processing

Beyer, E. et al. (Fraunhofer-Institute fur Lasertechnology Aachen, Germany). *Laser & Optoelektron.*, 17(3): 274-277, 281, 1985. (In German)

1.4 (42) The incredible laser – production tool

Williams, V. A. *Production*, 96(1): 41-47, 1985.

Lasers are being used to cut, weld, drill, mark and perform various metalworking operations.

Today's lasers are ruggedly constructed to withstand the production shopfloor environment, and offer improved reliability, self-diagnostics and simplified maintenance procedures.

1.4 (43) Industrial lasers and applications

Tuz, Z. L. (Laser Laboratory (Australia) Pty Ltd, Australia). *Mech. Engng. Trans. Inst. Engng.*, ME 9(3): 215-223, 1984. (Also in *Sheet Metal Industries*, 62(8): 418, 420-421, 1985.

Medium-power lasers are used in the medical field, and high-power lasers – from the 50W to 15kW range – are gaining rapid acceptance by the manufacturing industry. This paper presents an overview of the principles involved and their current use in metal cutting, welding, drilling and heat treatment.

1.4 (44) The laser as an industrial tool

Wautelet, M. and Laude, L. (Université de l'Etat, Mons, Belgium). *Recherche*, 16(169): 1026-1036, 1985.

By delivering its power in a very localised manner, the laser is capable of piercing, welding and cutting out with speed and precision. It has become effective in a number of uses from surface hardening to melting and doping of silicon.

1.4 (45) The laser does it all

Thorn, R. *Metalwork. Prod.*, 129(8): 49-56, 1985.

Reviews the latest trends and future potential of lasers in industry, ranging over selective hardening with minimal distortion, and welding thin dissimilar metals from one side only. A future possibility could be to assist conventional machine tools by the laser melting of difficult materials ahead of the tool. Other functions dealt with cover changing surface microstructures and surface alloying and cladding.

1.4 (46) Laser material processing

Nagai, H. (Mitsubishi Electric Corp, Japan) *J. Jpn. Inst. Electron. & Commun. Eng.*, 68(4): 412-422, 1985.

Describes the development of high-pressure sealed CW CO_2 laser with high efficiency for practical use in material processing. With improved levels of technology in terms of operating performance, functions, reliability, stability, etc. and shortened product life cycle along with diversified user needs, the laser material processor is now regarded as a general-purpose machine tool. The characteristics, application fields and the structure of YAG laser and CO_2 laser material processing are described.

1.4 (47) Laser systems for materials processing (CO₂ and solid state lasers)

Loosen, P. et al. (Technical Hochschule, Germany). *Feinwerktech. & Messtech.* 93(5): 222-233, 1985. (In German)

Describes CO_2 and solid state lasers for which the most important fields of application are cutting drilling and welding. Equipment for the heat treatment and refining of metals is generally still at the experimental stage and hardening of non-metals and paint coatings can only be achieved in isolated cases. Includes tables giving a market survey of the lasers supplied, the processing equipment and the manufacturers and sales companies involved. (9 refs.)

1.4 (48) Lightweight lasers for industry

Dennis, R. B. and Herman, H. *Mater. Design*, 6(1): 30-32, 1985.

Describes the low power CO_2 laser and its growing industrial applications in cutting drilling, soldering and marking.

1.4 (49) The management of industrial lasers

Green, B. G. and Bragg, M. J. (Laser Scientific Services Ltd, UK). In, *Proc. 2nd Int. Conf. on Lasers in Manufacturing*, 26-28 March 1985, Birmingham, UK, (Ed. M. F. Kimmitt), pp. 285-294. IFS (Publications) Ltd, Bedford, UK, 1985.

This paper, of particular interest to industrialists and commercial laser operators, covers the full spectrum of subjects relevant to the management of industrial lasers in 1985. Equipment aspects covered are identification of requirement; how to procure a laser system; the advantages of purchasing from a supplier with sub-contract facilities; pitfalls; staff training; day-to-day running and quality assurance. A financial appraisal shows that an industrial laser can be operated profitably without difficulty, particularly where high utilisation is achieved.

1.4 (50) Possible application of laser techniques

Behnisch, H. *Werstatt & Betr.*, 118(3): 169-172, 1985 (In German)

Welding Institute built fast axial flow carbon dioxide laser (maximum rated power 2.5kW) (See 1.4 (52))

Laser techniques are particularly valuable for welding in precision and micro-engineering and for cutting foil and sheet metal. The process consists of converting the energy of coherent light beam into useful heat. The coherent light of defined wavelength is focused so that heat is generated when it strikes a workpiece. This usually causes the metal to melt or vapourise. The width and depth of molten metal depend on the amount of laser energy and the absorption behaviour of the material relative to laser light of a certain wavelength. Neither electrical contact nor material contact with the heat source is necessary. Preferably no filler material is used. The big advantage of this process is that high precision joints can be made with a minimum heat input.

1.4 (51) The use of laser beams for welding, cutting, surface hardening, etc. – state of the art

Behnisch, H. *Laser & Optoelektron.*, 17(4): 385-393, 1985. (In German)

1.4 (52) The use of lasers in manufacturing-relevant research at the Welding Institute

Oakley, P. J. et al. (The Welding Institute, UK). In, *Proc. 2nd Int. Conf. on Lasers in Manufacturing*, 26-28 March 1985, Birmingham, UK, (Ed. M. F. Kimmitt), pp. 237-248. IFS (Publications) Ltd, Bedford, UK, 1985.

Briefly reviews the development of CO_2 laser equipments, and then describes typical results from process research work with Nd:YAG and CO_2 lasers relating to microjoining, welding of sheet and plate, and to surfacing.

1.4 (53) Laser activities in Japan

Fujioka, T. (IRI Laser Laboratory, Industrial Research Institute, Japan). In, *Proc. 3rd Int. Conf. on Lasers in Manufacturing*, 3-5 June 1986, Paris, France, (Ed. A. Quenzer), pp. 11-17. IFS (Publications) Ltd, Bedford, UK, 1986.

Describes the national projects related to lasers, the organisations conducting laser research, total sales of laser systems, and the future prospects in Japan.

1.4 (54) Laser material processing in the United States: a 1986 overview

Albright, C. (Ohio State University, USA). In, *Proc. 3rd Int. Conf. on Lasers in Manufacturing*, 3-5 June 1986, Paris, France, (Ed. A. Quenzer), pp. 5-10. IFS (Publications) Ltd, Bedford, UK, 1986.

Laser material processing in the United States is in an 'adolescent' stage, not yet mature, but far from infancy. Laser welding is a very rapid growing technology in the USA. New concerns over carcinogenic materials generated in the processing of polymers are arising. The Edison Welding Institute has been established and is sponsoring laser welding research.

2

LASER SYSTEMS

2.1 TYPES AND OPTICAL EQUIPMENT

2.1 (1) Optical design problems in industrial solid state lasers

Koechner, W. and Sooy, W. R. (Hadron Inc., USA). In, *Proc. SPIE Conf. on Optical Design Problems in Laser Systems*, 21-22 August 1975, San Diego, California, Vol. 69, pp. 14-20. Society of Photo-Optical Instrumentation Engineers, Palos Verdes Estates, CA, USA, 1975.

Discusses the optical peripherals, which include appropriate optics for shaping, directing and focusing the laser beam, optics for viewing of the workpiece, components to protect optics from contamination due to ejected material, beam monitors, radiation enclosures and safety shields necessary to provide for the safety of operating personnel.

2.1 (2) Laser welding and cutting systems

Anderson, D. G. (Union Carbide Corp, USA). In, *Proc. Latest Uses of Lasers in Manufacturing Conf.* 30 November – 2 December 1976, Culver City, CA. (SME Tech. Paper No. MR76-85). Society of Manufacturing Engineers, Dearborn, MI, USA, 1976.

High speed, reliable operation and quality of finished product are some of the advantages of CO_2 laser metalworking and these are emphasised in this paper. Also includes a discussion of the systems approach to laser metalworking, the advantages and limitations of laser welding and cutting, and a number of industrial welding and cutting applications.

2.1 (3) Tooling up for laser welding

Engel, S. L. (SME Tech. Paper No. MR76-873), 13 pp., 1976. Society of Manufacturing Engineers, Dearborn, MI, USA.

2.1 (4) Multikilowatt laser processing

Megaw, J. H. P. C. and Kaye, A. S. (UKAEA, UK). In, *Proc. Laser 77 3rd Opto-Electronics Conf.*, 20 – 24 June 1977, Munich, pp. 291-298. IPC Science and Technology Press, Guildford, UK, 1977.

A system using gas-transport CO_2 lasers designed and built at the UKAEA Culham Laboratory is described. An account is given of the characteristics of deep penetration laser welding, with particular reference to plasma effects and beam coupling efficiency; penetration capabilities are noted, and sample welds are shown. Also discusses investigations and results of transformation hardening and surface alloying by laser.

2.1 (5) Experiments on multiplane balancing using a laser for material removal

Demuth, R. S. (Mechanical Technology Inc., USA). NASA Contract Report No. 3105, pp. 1-40, 1979.

Experimental testing of the laser material removal method for balancing through the first bending critical speed was demonstrated. The rotor test rig, optical configuration, and a neodymium glass laser system were assembled and calibrated for static and rotating material removal rates. The laser control computer program was combined with the influence coefficient balancing process, resulting in a completely automated data acquisition, laser, and balancing system.

2.1 (6) High-power laser materials processing

Banas, C. M. (United Technology Research Center, USA). In, *Proc. 2nd Gas-Flow and Chemical Lasers Int. Symp.*, 1978, Rhode-St-Genese, Belgium, pp. 37-48. Hemisphere Publishing Corp., Washington, DC, 1979.

Reviews the materials processing capabilities

of high-power laser systems. Operating regimes for laser shock hardening, drilling, welding, and transformation hardening are identified and processes described. Unique characteristics of laser processing are discussed, and typical performance and process limits are noted. (5 refs.)

2.1 (7) Positioning system with laser for automatic micro spot laser welding

Ono, A. et al. *Ann. CIRP*, 28(1): 317-320, 1979.

A positioning system using a He-Ne laser was developed for automatic micro spot YAG laser welding. Positioning was attained by detection of the reflection beam from the edge of the part close to the welding spot with the He-Ne laser beam scanned over ·the edge. The angle of incidence to the axis of the part is selected to reduce disturbance light reflection. After detecting the edge, the YAG laser welding optical system moves a given distance from the edge to the accurate welding spot, which it then irradiates. This YAG laser welder is used in automatic assembly systems for mass production of high quality electric devices.

2.1 (8) Automatic workpiece surface tracker for laser cutter

Mottier, F. M. (United Technology Corp, USA). In, *Proc. SPIE Conf. on Optics in Metrology and Quality Assurance*, Vol. 220, 6-7 February 1980, Los Angeles, pp. 95-100. Society of Photo-Optical Instrumentation Engineers, Bellingham, WA, USA, 1980.

Precise guidance of high energy laser beams is important in industrial laser machining to optimise the use of the available power and to assure high-quality work. Describes a crossed beam depth gauge that measures the distance to the workpiece surface with an accuracy of one part in two thousand over a range of 10 cm.

2.1 (9) Debris pick-up and lens shield for laser machines

Townsend, T. A. (Western Electric, USA). *Tech. Digest*, (59):27-28, July 1980.

In the laser machining of moving parts, such as a ceramic thin film substrate, a focused laser beam is transmitted from a lens of the laser head a short distance to the surface of the substrate to be machined. The substrate is mounted on a table and moved rapidly under the beam to machine the part. A problem is that, with the substrate moving rapidly under the beam, a high-speed stream of laser debris is ejected from the surface, consisting of minute particles of the disintegrated ceramic. A specially contoured

exhaust housing is described. It is positioned between the lens and the substrate, and functions to pick up the debris and to protect the lens.

2.1 (10) Laser fabrication machine failure detector

Belyeu, S. M. (IBM Corp, USA). *IBM Tech. Disclosure Bulletin*, 23(3): 883-886, 1980.

A failure detector is described which senses laser shutter and worktable movement failures during the operation of laser fabrication machines, such as integrated circuit chip personalisation machines, laser cutting and drilling machines. Describes a typical machine and the details of a failure detector are given.

2.1 (11) Characteristic features of the inert-gas shielding of the heated area during laser welding and heat treatment

Cherkashin, A. P. et al. *Svar. Proizvod.*, (12): 19-20, 1982. (In Russian)

An experiment has been used to optimise the nozzle design and shielding conditions. As a result, a general-purpose inert gas nozzle has been developed for laser welding and heat treatment. The schematic design of the nozzle is presented, along with its performance characteristics.

2.1 (12) Economy and reliability of laser cutting machines

Balbach, J. and Tonshoff, H. K. *Bander Bleche Rohre*, 23(10): 294-296, 1982. (In German)

High capital costs are countered by low operating costs. Combined stamping and laser-cutting machines offer a minimum holding time for the workpiece which can be processed in a single chuck to yield complex outline configurations and simple perforations. Safety measures to be implemented are set by a classification of the laser according to limits on the radiation which replaces the difficult and costly measurement of irradiation values.

2.1 (13) Laser flash inhibitor for laser monitoring and camera protection

Holt, R. D. (Bendix Corp, USA). Rept. No: BDX-613-2795, 18 pp., July 1982. Bendix Corp, Kansas City, MO, USA.

Describes the development of a fast electronic shutter in a high powered Nd:YAG laser. This system eases visual in-process monitoring and protection for the camera's vidicon tube. The simplicity and great benefits in process monitoring have made the use of the laser flash inhibitor very successful and significantly improved the Bendix capability to use the laser welding system.

2.1 (14) Special features of inert gas shielding the heating area in laser welding and heat treatment

Mironov, L. G. et al. (Scientific and Production Organisation Kuant, USSR). *Weld. Prod.*, 29(12): 14-16, 1982.

Chromatographic methods were used. A universal nozzle for supplying the gas to laser welding equipment was designed. The conditions of shielding the heated area with the universal nozzle and inert gas were determined.

2.1 (15) Cutting and welding three-dimensional sheet metal parts with high accuracy

Nilsson, K. and Sarady, I. (University of Lulea, Sweden). In, *Proc. 1st Int. Conf. on Lasers in Manufacturing*, 1-3 November 1983, Brighton, UK, (Ed. M. F. Kimmitt) pp. 79-86. IFS (Publications) Ltd, Bedford, UK, 1983.

By laser cutting followed by laser butt welding, three-dimensional sheet metal parts can be manufactured, replacing resistance spot-welded lap joints. For this purpose two lasers have been integrated to form a 3- to 5-axis NC coordinate system. The system has adaptive compensation for the proper positioning of the focal point on the surface of the object or workpiece. Another adaptive compensation is being developed for seam tracking. The soundness of the resulting welds has been affirmed by tensile and fatigue testing.

2.1 (16) Experimental studies in nozzle design for laser cutting

Thomassen, F. B. and Olsen, F. O. (Technical University of Denmark). In, *Proc. 1st Int. Conf. on Lasers in Manufacturing*, 1-3 November 1983, Brighton, UK (Ed. M. F. Kimmit), pp. 169-180. IFS (Publications) Ltd, Bedford, UK, 1983.

Based on about 500 experiments, this paper analyses the importance of nozzle pressure and cutting speed for the quality of the cut. The experiments are made with three different types of nozzles, in mild steel and in stainless steel. In mild steel a higher nozzle pressure, until a certain level, increases the rates of high quality cuttings. At pressures above this level some small burning marks appear in the kerf. Self burning marks like these are not seen when

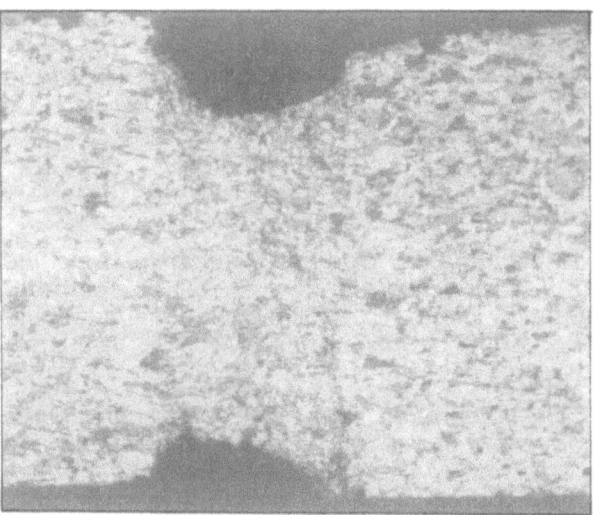

Butt weld with He as shield gas (See 2.1 (15))

cutting in stainless steel within our maximum pressure. Parameters like focal spot size and polarisation are examined. The striations occuring in the kerf are analysed and compared with the sparks appearing.

2.1 (17) Metal processing at Culham

Spalding, I. J. In, *Proc. Physical Processes in Laser/Materials Interactions*, 13-25 July 1980, Pianore, Italy, pp. 271-282. Plenum Press, New York, 1983.

The Laser Applications Group at Culham utilises a wide range of commercially-available and in-house CO_2 lasers, covering the power range 5-15,000W (continuous). The subkilowatt systems utilise stable optical resonators, providing Gaussian-mode outputs, and some of these have high mean-power pulsing facilities at repetition rates up to 1kHz. Pressure vessel and tube which are to become tube-sheet laser weldments, are illustrated.

2.1 (18) A c-radius technique for determination of beam profiles

Chang, D. U. (Ford Motor Company, USA). In, *ICALEO '83, Proc. Materials Processing Symposium*, Vol. 38, 14-17 November 1983, Los Angeles, CA, (Ed. E. A. Metzbower), pp. 22-30. Laser Institute of America, Toledo, OH, USA, 1984.

A simple technique for determination of the beam profile is described. This technique is based on the c-radius concept, and does not require elaborate equipment.
An example of beam profile determination of a CO_2 laser is presented. A heat-sensitive thermal printer paper was used to generate the images of the laser beam at various exposure times.

Based on the images, the beam profile was plotted and compared with a Gaussian profile. This technique also allows an estimation of the $1/e^2$ radius of a near Gaussian beam. (9 refs.)

2.1 (19) Computer-aided engineering of a laser machining center

Krutilla, M. A. In, *Proc. 2nd Biennial Int. Machine Tool Technical Conf.*, Vol. 3, 5-13 September 1984, Chicago, pp. 8/65-84. National Machine Tool Builders Association, McLean, VA, USA, 1984.

The integration of the CAE systems into the conventional design process is described using a laser machining centre as an example, which was designed, analysed, and built in a period of five months.

2.1 (20) Flexible beam delivery for material processing laser power through a fiber optic cable

Georgalas, G. and Jones, M. G. (GEC, USA). In, *ICALEO '83, Proc. Materials Processing Symp.*, Vol. 38, 14-17 November 1983, Los Angeles, (Ed. E. A. Metzbower), pp. 149-153. Laser Institute of America, Toledo, OH, USA, 1984.

A neodymium: yttrium-aluminum-garnet (Nd:YAG) laser used in a pulsed mode is coupled to and passed through a single quartz fibre optic cable at peak power levels required for laser material processing. This flexible fibre permits laser cutting, drilling, and welding of metals with a robot. Metals in excess of 2mm (0.080 inches) thick have been laser cut.

2.1 (21) Hemispherical reflector for laser processing

Goodman, D. S. *IBM Tech. Discl. Bulletin*, 27(5): 31-41, 1984.

The reflector allows a more efficient use of laser light in processing and greater safety.

2.1 (22) Laser beam diagnostics for metal working applications

Ramos, T. J. and Lingenfelter, R. C. (Lawerence Livermore National Laboratory, USA). In, *Proc. Int. Conf. on Applied Lasers and Electro-Optics*, 12 November 1984, Boston, MA, report No. UCRL-90805, CONF-8411100-5.

Describes the Materials Fabrication Division (MFD) of Lawrence Livermore National Laboratory (LLNL) pulsed Nd:YAG lasers dedicated to metal working. The units are used for welding, brazing, cutting, and drilling. The MFD facility is a job shop fabricating prototypes of design

concepts. Reviews the procedure used for set-up and operation of the laser metal working systems, with particular emphasis on the methods used to define and control the mode, stability and symmetry of the laser beam.

2.1 (23) Nd: YAG laser improvements and modifications

Holt, R. D. and Mallory, M. B. (Bendix Corp, USA). Rept. No. BDX-613-2911, 31pp., March 1984. Bendix Corp, Kansas City, MO, USA.

Many improvements and additions were made, including output power stability, flash lamp life, a laser flash inhibitor, a rotary axis, wire feed, power-driven camera focus, and a power-driven laser lens positioning system with digital display of position. The laser system functions fully and reliably on the production line.

2.1 (24) Where laser cutting is of interest

Aerni, G. *Technica*, 33(12): 30-32, 1984. (In German)

A medium size factory in Switzerland producing large and small scale factory buildings, staircases, window frames, etc. uses a computer controlled precision CO_2 laser for metal cutting. The machine is capable of cutting wood, synthetic materials, chromium steel 4301/4401, St 37 steel, etc. The gas mixture consists of He, Ne, CO_2 and is prepared automatically; the machine is rated at 600W. The laser cutter proved most valuable when cutting sharp corners, curves, pointed pieces, thin strips, etc.

2.1 (25) The CO_2 waveguide laser (a laser with designs on industry)

Ross, I. E. (Ferranti plc, UK). In, *Proc. 2nd Int. Conf., on Lasers in Manufacturing*, 26-28 March 1985, Birmingham, UK, (Ed. M. F. Kimmitt), pp. 209-218. IFS (Publications) Ltd, Bedford, UK, 1985.

The CO_2 waveguide laser has developed from military laser technology which has recently proven itself as a tool for manufacturing industry. The Laser is based on rugged ceramic technology, has a sealed long-life gas fill, giving excellent MTBF, and is extremely controllable with on/off switching and pulsing on command.

The development of the technology, the specific advantages in performance and a taste of the possible applications will be described.

A 20W CO₂ waveguide laser (See 2.1 (25))

2.1 (26) Development of metal mirrors for high power CO₂ lasers

Sumiya, M., Ueda, K. (Toshiba Corp, Japan) and Kawata, K. (Matsushita Research Institute, Japan). In, *ICALEO '84, Proc. Materials Processing Symp.*, Vol. 44, 12-15 November 1984, Boston, MA, (Ed. J. Mazumder), pp. 276-283. Laser Institute of America, Toledo, OH, USA, 1985.

This paper presents surface finishing of metal mirrors for high-power CO₂ laser systems. A diamond turning machine has been developed to make copper mirrors. The machined mirrors have high contour accuracy and excellent surface roughness. Mechano-Chemical polishing of molybdenum and tungsten, and the optical properties of the surface are also discussed.

2.1 (27) Designing the 'industrial strength' excimer. III

Van Scoy, R. L. *Photonics Spectra*, 19(6): 53-56, 1985.

(For pt. II see ibid., 19(5): 67-70, 1985.) This, part considers the transferring of the excimer laser from the laboratory bench to the factory floor, concentrating on those areas and applications that make the excimer workstation a desirable industrial tool.

2.1 (28) A 5kW CW CO₂ laser for industrial applications

Fantini, V. et al. (CISE SpA, Italy). In, *Gas Flow and Chemical Lasers, 1984, Proc. 5th Int. Symp.*, Adam Hilger, Bristol, UK, 1985, 534pp.

It is a transverse flow, electrical discharge laser. It has been designed with two independent discharge channels and can operate as a 5kW laser source, using the two channels together, or as two 2.5kW independent laser sources, each one using one channel. Some results of laser welding and heat treatment experiments are also shown.

2.1 (29) High-power electroionization CO₂ and CO lasers for industrial applications

Basov, N. G. et al. (P. N. Lebedev Physics Institute, USSR). *IEEE J. Quantum Electron.* QE-21(4): 342-358, 1985.

Research conducted in the USSR over the past decade on lasers for industrial applications is discussed. A laser system which can run on multiple wavelengths, can operate in CW and long or short repetitive pulse modes, and can operate economically was the design goal. The result is a laser system which uses a coupled turbocompressor and refrigerator to supply the gas flow and temperature control. The system can operate on CO or CO₂, thus serving as a source of mid-infrared and far-infrared radiation in either the CW or pulse repetition mode.

2.1 (30) Laser beam diagnostics for metal working applications

Ramos, T. J. and Lingenfelter, A. C. (Lawrence Livermore National Laboratory, USA). In, *ICALEO '84, Proc. Materials Processing Symp.*, Vol. 44, 12-15 November 1984, Boston, MA, (Ed. J. Mazumder), pp. 306-315. Laser Institute of America, Toledo, OH, USA, 1985.

Description of the laboratory's pulsed Nd:YAG lasers for welding, brazing, cutting and drilling. Emphasises the mode symmetry and stability of the beam.

A reflective coupler is described for materials processing applications with infrared lasers. A hemispherical integrating cavity with a conical lower section is used to increase the absorptivity and the emissivity of the surface to be treated, while at the same time producing a more uniform and sharply edged heat distribution. (14 refs.)

2.1 (31) A laser-surface coupler for efficient materials processing

Lamonde, G. (Gentec Inc, Canada) and Cielo, P. (Industrial Materials Research Institute, National Research Council of Canada). In, *ICALEO '84, Proc. Materials Processing Symp*, Vol. 44, 12-15 November 1984, Boston, MA, (Ed. J. Mazumder), pp. 298-305. Laser Institute of America, Toledo, OH, 1985.

2.1 (32) New CO$_2$ lasers of 1-4 kilowatt power with fast axial gas flow

Bakowsky, L. (Messer Griesheim GmbH, West Germany). In, *Proc. 2nd Int. Conf. on Lasers in Manufacturing*, 26-28 March 1985, Birmingham, UK, (Ed. M. F. Kimmitt), pp. 195-200. IFS (Publications) Ltd, Bedford, UK, 1985.

Messer Griesheim has developed a new series of high-performance CO$_2$ gas lasers for materials processing applications (EUROLAS). This series comprises a modular system which includes four performance classes at 1kW intervals from

Resonator of the EUROLAS 1000 laser system (See 2.1 (32))

1kW up to 4kW. All models feature simple, robust design, high mode and power stability, and high specific laser power (1kW/m).

Because of the high specific power it is possible to get 1kW laser from only two discharge tubes of 500mm length each. The laser can be operated both continuously (CW) and pulsed.

2.1 (33) The start of a new generation of CO$_2$ lasers for industry

Hoffmann, P. (Laser Innovation GmbH & Co KG, West Germany). In, *Proc. 2nd Int. Conf. on Lasers in Manufacturing*, 26-28 March 1985, Birmingham, UK, (Ed. M. F. Kimmitt), pp. 201-208. IFS (Publications) Ltd, Bedford, UK, 1985.

The present, first generation of industrial CO$_2$ lasers originated in the period 1966 – 1970 and is differentiated into axial flow-axial discharge and transverse flow-transverse discharge types, all using DC glow discharges to generate the required population inversions. Describes one of the first of the new, second generation of devices, utilising an RF-excited transverse flow-transverse glow discharge, 3-axis configuration which provides greatly increased power density in the discharge plasma. This approach permits significant reductions in both head and power supply sizes and weights, simplicity of construction, and reduced manufacturing costs and maintenance requirements.

2.1 (34) Supersonic characteristics of nozzles used with lasers for cutting

Ward, B. A. (Laser Applications Group, Culham Laboratory UKAEA, UK). In *ICALEO '84, Proc. Materials Processing Symp.*, Vol. 44, 12-15 November 1984, Boston, MA, (Ed. J. Mazumder), pp. 94-101. Laser Institute of America, Toledo, OH, USA, 1985.

Many operators of laser cutting equipment use supersonic flow of a cut-assisting gas from a nozzle directed at the workpiece. This flow gives rise to shock-fronts which can cause large, localised variations in the effective gas pressure on the workpiece. This work has used pressure measurement techniques to investigate the variations in air-jet pressures. It is possible to identify the optimum value of parameters that will result in a high pressure on the workpiece. CO$_2$ laser cutting trials demonstrate the strong correlation between the workpiece pressure and maximum cutting speed. The pressure measurements reveal the sensitivity of the flow pattern to nozzle damage and manufacturing variations. A simple nozzle 'tool-setting' and damage inspection technique is proposed. This technique is compatible with an industrial production environment.

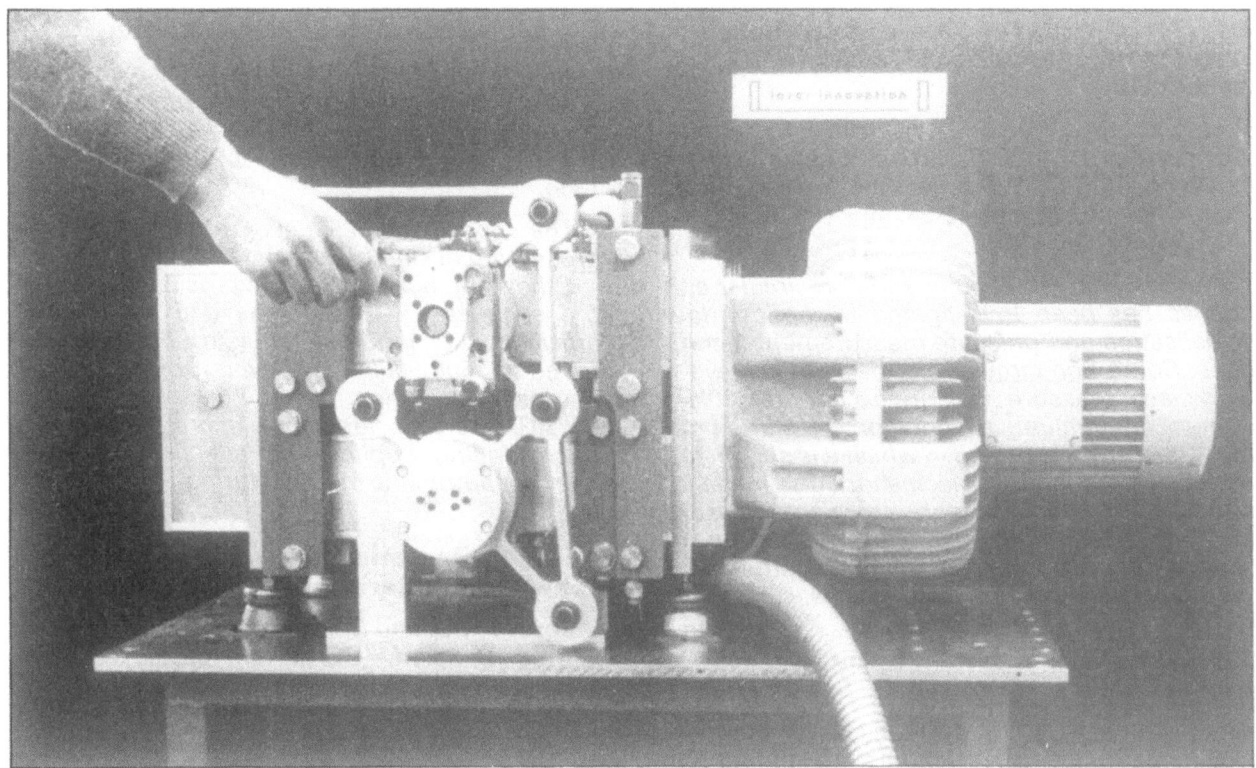

The prototype laser head (See 2.1 (33))

2.1 (35) 2kW jobbing laser

Hardisty, F. B. (Bramah Engineering, UK). *Met. Constr.* 17(2): 87-89, 1985.

An account is given of how Bramah Engineering decided to purchase a 2kW CO_2 laser for use in jobbing type applications and how the system was implemented for laser cutting, welding and surface treatments.

2.1 (36) Adaptive control for high speed and high quality laser cutting

Moriyasu, M. et al. (Mitsubishi Electric Corp, Japan). In *Laser Welding, Machining and Materials Processing – Proc. Int. Conf. on Applications of Lasers and Electro-optics, ICALEO '85*, 11-14 November 1985, San Francisco, CA, (Ed. C. Albright), pp. 129-136. Laser Institute of America, Toledo, OH, USA/IFS (Publications) Ltd, Bedford, UK, 1986.

In the case of a complex figure cut with CO_2 laser, actual cutting speed changes momentarily at corner with directional conversion. For the purpose of high-quality cutting, it is necessary for such parameters as laser power, laser output mode and cutting speed to be kept appropriate.

The new control system for CO_2 laser cutting has been developed, where laser cutting parameters are controlled adaptively with a mutual relation between laser power, output mode and cutting speed in order to ensure high efficiency and high quality.

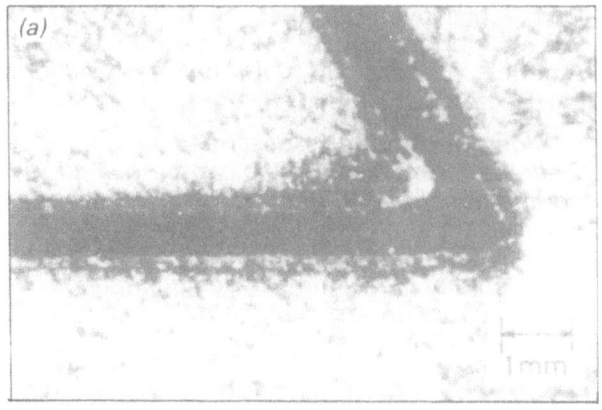

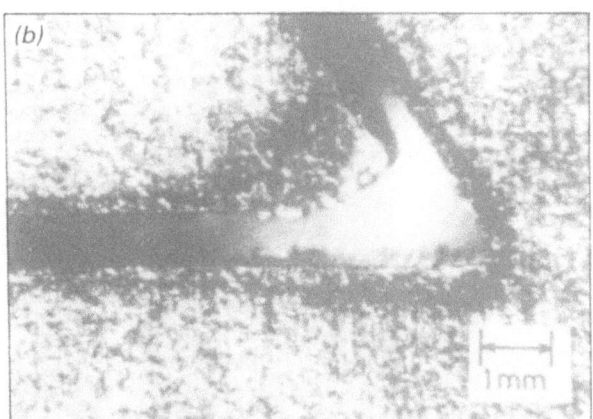

Effect of adaptive control: (a) with adaptive control (5mm/min), and (b) without adaptive control (CW 500W, 5mm/min) (See 2.1 (36))

2.1 (37) Circular and non-circular nozzle exits for supersonic gas jet assist in CO₂ laser cutting

Fieret, J. and Ward, B. A. (UKAEA Culham Laboratory, UK). In, *Proc. 3rd Int. Conf. on Lasers in Manufacturing*, 3-5 June 1986, Paris, France, (Ed. A. Quenzer), pp. 45-54. IFS (Publications) Ltd, Bedford, UK, 1986.

A wide variety of circular and non-circular exit nozzles has been investigated by optical flow visualisation and pressure measurement techniques. Some non-circular exit designs achieve a workpiece pressure well in excess of 600kPa (gauge). A qualitative supersonic jet impingement model is presented in which it is suggested that the formation of a Mach shock disc in the jet must be avoided for a high workpiece pressure at practical working distance. The design of suggested nozzle exits is given, together with the associated pressure distributions.

2.1 (38) In-process monitoring of laser processes

Weerasinghe, V. M. and Steen, W. M. (Imperial College, UK). In, *Laser Welding, Machining and Materials Processing – Proc. Int. Conf. on Applications of Lasers and Electro-Optics, ICALEO '85*, 11-14 November 1985, San Francisco, CA, (Ed. C. Albright), pp. 107-112. Laser Institute of America, Toledo, OH, USA/IFS (Publications) Ltd, Bedford, UK, 1986.

A new technique is described in this paper whereby in-process signals can be gained with no additional beam interference. The technique is based on analysing the acoustic signals from the mirrors reflecting high powered laser beams. It is seen that these mirrors are ringing under the photon stress imposed on them.

2.1 (39) Laser processing and products

Clement, P. (Limoges Precison, France). In, *Proc. 3rd Int. Conf. on Lasers in Manufacturing*, 3-5 June 1986, Paris, France, (Ed. A. Quenzer), pp. 337-339. IFS (Publications) Ltd, Bedford, UK, 1986.

The report mainly deals with the CO₂ laser processing of sheet-form materials.
Other high power laser systems used in industry, such as YAG lasers, have a much lower efficiency than gas lasers, maximum 400W compared to 1500-5000W and even more in the new gas laser generation.
YAG lasers designed for micro-welding and micro-drilling applications are practically not used for automatic manufacturing.

2.1 (40) Operation of a multi-station 25kW laser material processing facility

Gregson, V. G. and Morgan, D. F. (Westinghouse Marine Division, USA). In, *Laser Welding, Machining and Materials Processing – Proc. Int. Conf. on Applications of Lasers and Electro-Optics, ICALEO '85*, 11-14 November 1985, San Francisco, CA, (Ed. C. Albright) pp. 163-168. Laser Institute of America, Toledo, OH, USA/IFS (Publications) Ltd, Bedford, UK, 1986.

The high power laser processing facility has achieved the reliability, and applications goals expected. However, rapid change-over from one job lot to another rather than automation is the next goal.

2.1 (41) RF-excitation for high power CO₂ lasers

Schock, W. (DFVLR, Institut für Technische Phy-

Etched section of a laser processed ½in. stainless steel joint (See 2.1 (40))

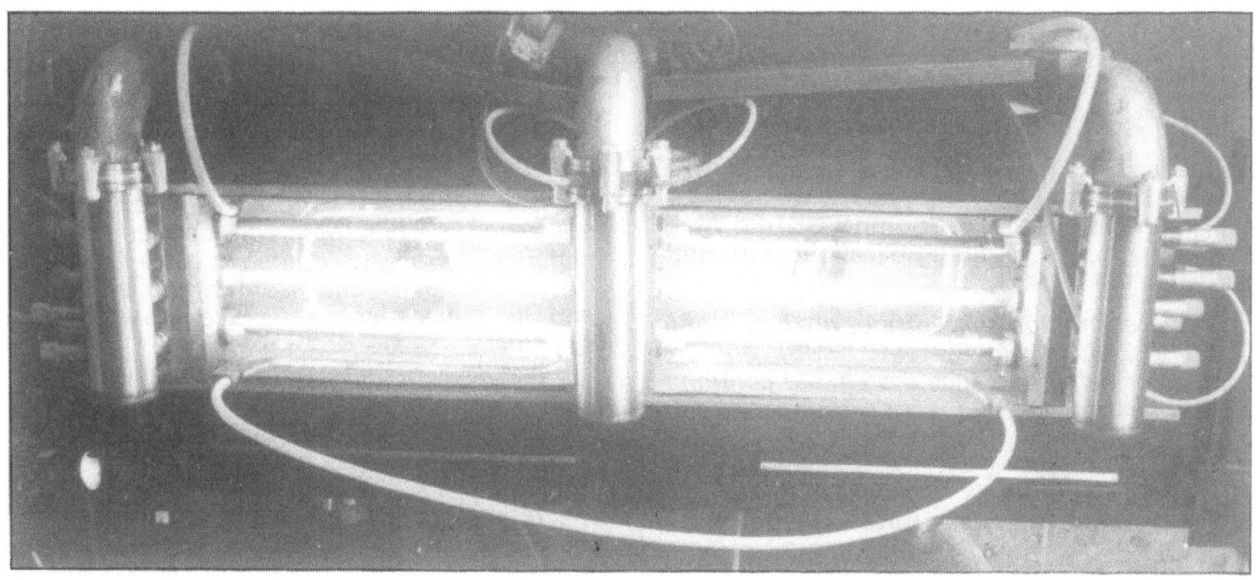

Fast axial flow laser in operation (See 2.1 (41))

sik, West Germany). In, *Proc. 3rd Int. Conf. on Lasers in Manufacturing*, 3-5 June 1986, Paris, France, (Ed. A. Quenzer), pp. 271-278. IFS (Publications) Ltd, Bedford, UK, 1986.

Experimental investigations with both transverse and axial flow CO_2 lasers demonstrate the outstanding potential of radio-frequency discharges for the excitation of modern high-power lasers. In particular, optical power densities of up to 5W/cm^3 are achieved at near diffraction limited beam quality, and CW as well as pulsed mode operation is possible for a wide range of discharge parameters.

2.1 (42) RF-excited high power CO_2 lasers for industrial material processing

Wollermann-Windgasse, R. (TRUMPF GmbH & Co West Germany). In, *Proc. 3rd Int. Conf. on Lasers in Manufacturing*, 3-5 June 1986, Paris, France, (Ed. A. Quenzer), pp. 293-304. IFS (Publications) Ltd, Bedford, UK, 1986.

Industrial CO_2 lasers excited by an RF-discharge are presented. The lasers are of the fast axial flow type with maximum output powers of 1000W and 1500W CW. Fast power variation is possible between 5% and 100% of maximum output retaining a very stable discharge in the whole power range. Pulsing of the lasers is done by switching the RF-discharge with repetition rates up to 10kHz and pulse durations from approximately 100μs on. By these features the lasers can be adjusted very well to the process requirements. Integration and working results in flexible manufacturing systems are described. Within this the technical features of a 5-axis laser cutting machine with superb path and positioning accuracy are presented.

2.1 (43) Sensory characteristics of the 'Melcut-3DCM'

Sibayama, K., Kubo, M. and Itani, K. (Mitsubishi Electric Corporation, Japan). In, *Laser Welding, Machining and Materials Processing, Proc. Int. Conf. on Application of Lasers and Electro-Optics, ICALEO '85,* 11-14 November 1985, San Francisco, CA, (Ed. C. Albright), pp. 113-120. Laser Institute of America, Toledo, OH, USA/IFS (Publications) Ltd, Bedford, UK, 1986.

The three dimensional laser cutting machine 'MELCUT-3DCM' has a height sensor. With the aid of the height sensor, the operator can make a teaching program easily and any complex preformed metal sheet can be finely cut.
The height sensor has also a function of preventing misirradiation of CO_2 laser beam.

The Melcut-3DCM (See 2.1 (43))

2.2 LASERS AND ROBOTS

2.2 (1) Robot controlled laser processing

Johnson, T. A. (Flexible Laser Systems Ltd, UK). In, *Proc. 1st Int. Conf. on Lasers in Manufacturing*, 1-3 November 1983, Brighton, UK, (Ed. M. F. Kimmitt), pp. 71-78. IFS (Publications) Ltd, Bedford, UK, 1983.

This presentation concentrates on the Flexible Laser Manufacturing System (FLMS), and examines in detail one such system – a combination of beam guide and robotic assembly (COBRA).

2.2 (2) Welding with robots – Is it for you?

Quinlan, J. C. *Tool and Production*, 48(12): 93-97, 100, 1983.

Reports an update on robotic welding technology with a summary of its potential benefits. Expansion in the spot welding robot market is discussed. The market for arc-welding robots remains virtually untapped but the present status and available systems are discussed. Applications include spot welding automobile bodies, shopping carts and laser welding of plates for military armour.

2.2 (3) Laser welding technology

Hendrixson, D. (Cincinnati Milacron, USA). In, *Proc. 2nd Annual Int. Robot Conf, 1984*, pp. 305-332. Tower Management Co, Wheaton, IL, USA, 1984.

Laser technology coupled with robotics has unleashed a totally new and innovative manufacturing process. Outlines the development trends in robotic laser welding technology and its application in flexible manufacturing systems.

2.2 (4) Lasers make their mark

Engineer, 259 (6695): 32, 1984.

Briefly describes a robotic laser system for inscribing code-marks on curved diesel-engine injectors.

2.2 (5) Falcon/800 makes possible true laserobots

Eckersley, J. (Laser Corp of America, USA). In, *ICALEO '84, Proc. Materials Processing Symp.*, Vol. 44, 12-15 November 1984, Boston, MA, (Ed. J. Mazumder), pp. 193-197. Laser Institute of America, Toledo, OH, USA, 1985.

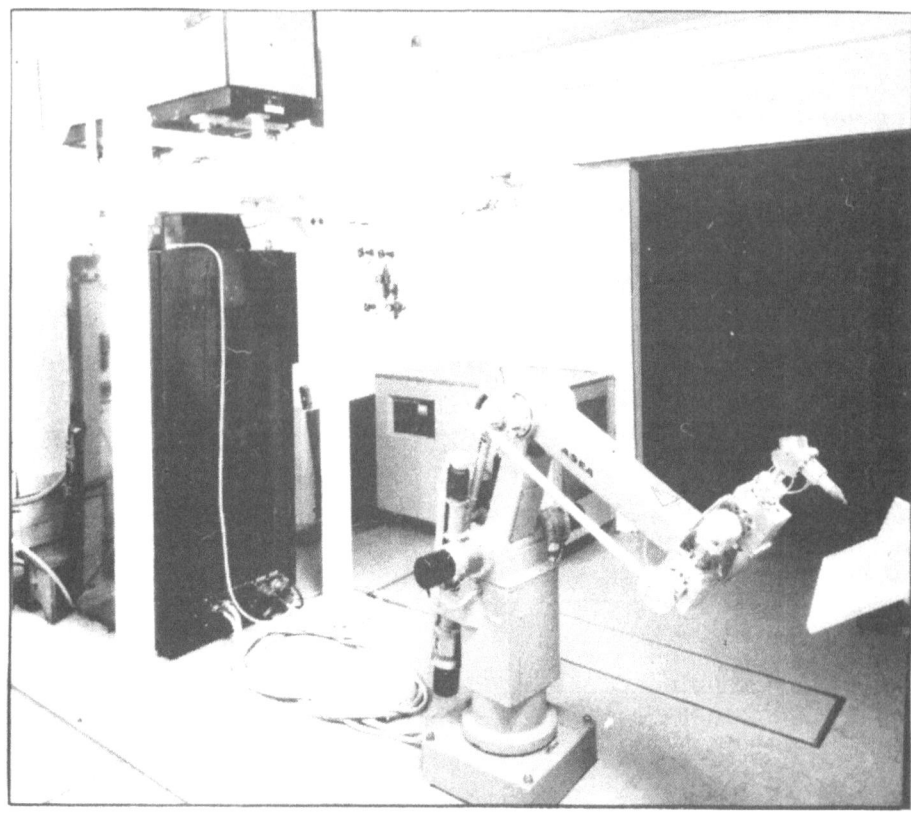

The complete COBRA system: in view are the laser and its control unit, the robot and program control unit and the flexible laser-beam guide connecting laser to cutting head (See 2.2 (1))

Brief description of this CO_2 laser which weighs 120lbs and can be mounted directly on to the robot manipulator.

2.2 (6) A laser robot for cutting and trimming deeply stamped metal sheets

Delle Piane, A. (Prima Progetti SpA, Italy). In, *Proc. 2nd Int. Conf. on Lasers in Manufacturing*, 26-28 March 1985, Birmingham, UK, (Ed. M. F. Kimmitt), pp. 219-224. IFS (Publications) Ltd, Bedford, UK, 1985.

Describes a robotised system, called ZAC, designed for cutting and trimming pre-formed metal sheets and deeply stamped plastic parts by means of a focused CO_2 power laser beam as a non-contacting tool.
The system allows the cutting of 3D pieces, moving and orienting the focused beam with 5 NC controlled movements (3 rectangular translations and 2 angular rotations).

2.2 (7) A rule based decision system for the robotisation of metal laser cutting

Foulloy, L., Kechemair, D. and Burg, B. (ETCA, France). In, *Proc. IEEE Int. Conf. on Robotics and Automation*, Cat. No. 85CH2152-7, 25-28 March 1985, St. Louis, MO, pp. 192-197. IEEE Computer Society Press, Silver Spring, MD, USA, 1985.

The conception of a rule based experimental system for the robotisation of metal cutting using a 5kW CO_2 continuous laser is considered. The image of the melted metal pulsed out of the kerf during the cut has been processed on the fly by a special-purpose vision system in order to estimate the cut quality. Methods used for the construction of the rules and the global architecture of the system are presented.

2.2 (8) Welding automation and robotics

Proceedings of a Conference, 12-13 March 1985, Indianapolis. American Welding Society, Miami, FL, 1985, 135pp.

The following topics were dealt with: economical robotic arc welding, including small batch manufacture; total concept of robotic manufacture; engine recuperator welding; rebuilding thin edges; gas tungsten arc welding; brazing and soldering; multipass welding; laser cutting; adaptive systems; robotics education; and tooling.

2.2 (9) How a man and a laser robot can learn mutually

Burg, B. (ETCA/CTME/OP, France). In, *Proc. 3rd Int. Conf. on Lasers in Manufacturing*, 3-5 June

Industrial robot with beam delivery system for a CO_2 laser (See 2.2 (10))

1986, Paris, France (Ed. A. Quenzer), pp. 89-96. IFS (Publications) Ltd, Bedford, UK, 1986.

High power continuous lasers are very flexible tools; unfortunately, their control in industrial applications is difficult. Thus, a rule based experimental device has been built at ETCA to evolve towards a robot enabled to control in real-time the various laser applications.

In its first version, the robot is occupied with metal laser cutting, by means of two main devices: a sensor incorporating a TV camera and a real-time image processor, and a rule based decision system to estimate the cut quality and to drive the process.

The robot is implemented as a very flexible tool, where the user can learr more about laser cutting, and then teach the robot in a quasi-natural language.

2.2 (10) Laser-fibre-robot system in high precision manufacturing

Roos, S.-O. (Permalux Lasersystem AB, Sweden). In, *Proc. 3rd Int. Conf. on Lasers in Manufacturing*, 3-5 June 1986, Paris, France, (Ed. A. Quenzer), pp. 321-327). IFS (Publications) Ltd, Bedford, UK, 1986.

The advantages of flexible laser machining are obtained first at the stage where you can combine the distribution of the radiation with robotics. In Nd laser processing the advantages of

POLYAS robot (See 2.2 (12))

using fibre optics together with robotics are obvious. The demands on the different parts of a laser-fibre-robot system are strongly dependent on the application but there are also general demands which can be anticipated. Parameters such as core diameters of fibres used and precision of the robot are of special interest when investigating different applications. This paper discusses applications in the fine mechanic and electronic field and compares the demands of accuracy with experimental results from a laser-fibre-robot system.

2.2 (11) Laser in high-rate industrial production automatic systems and laser robotics

Carroz, J. (Spectra-Physics, France). In, *Proc. 3rd Int. Conf. on Lasers in Manufacturing*, 3-5 June 1986, Paris, France, (Ed. A. Quenzer), pp. 345-354. IFS (Publications) Ltd, Bedford, UK, 1986.

Provides a general overview of laser systems running in high-rate industrial production. Description of laser systems for specific applications and various types of laser robotics is given with their production performances.

Gives an economical justification of such a system through a detailed estimation of the final cost of a laser welded part produced in high-rate industrial production.

2.2 (12) Laser robot for nuclear applications

Geffroy, J. et al. (CEA/IRDI/DEDR/DEMT, France). In, *Proc. 3rd Int. Conf. on Lasers in Manufacturing*, 3-5 June 1986, Paris, France, (Ed. A. Quenzer), pp. 313-320. IFS (Publications) Ltd, Bedford, UK, 1986.

Presents the laser tool interest in nuclear environment which is due specially to its ability to cut without mechanical stresses. Describes the constraints related to the use of these techniques in this environment. They lead to remote working. The focusing head of the laser tool must be moved in the three directions, either with the help of telemanipulators or robots, or directly by a robot itself. For this purpose, the authors have developed techniques of transmission of CO_2 power laser beam and have realised a prototype of a polyjointed robot with five rotation axis integrating the beam path in its structure (POLYAS robot). Describes this robot which constitutes a laser tool for industrial uses and serves as an experiment for the development of the laser techniques in nuclear environment.

2.2 (13) Review of laser-robotics-challenges remaining to fully develop the technology

Tight, T. (Spectra-Physics, USA). In, *Laser Welding, Machining and Materials Processing – Proc. Int. Conf. on Applications of Lasers and Electro-optics, ICALEO '85*, 11-14 November 1985, San Francisco, (Ed. C. Albright), pp. 95-100. Laser Institute of America, Toledo, OH, USA/IFS (Publications) Ltd, Bedford, UK, 1986.

Both industrial lasers and industrial robots have proven reliable, useful machine tools. The integration of these two technologies to form flexible workstations is a natural next step. This paper attempts to provide an overview of this new technology. The different approaches to combining lasers and robots are first described along with a listing of the 'pros' and 'cons' of each approach. The results of Spectra-Physics' development efforts in laser-robotics are briefly outlined, followed by a list of the current technical challenges for expanding the market to reach its full potential.

2.2 (14) Robolaser in the car industry

Aberman, Z. (Robomatix Ltd, Israel). In, *Proc. 3rd Int. Conf. on Lasers in Manufacturing*, 3-5 June 1986, Paris, France, (Ed. A. Quenzer) pp. 305-312. IFS (Publications) Ltd, Bedford, UK, 1986.

The robot laser which has been developed presents a flexible tool, enabling the opening of the required vents in the car body in the final stages of body manufacture, and goes into action with the designation of the vehicle.

The system provides a saving in the number of superfluous openings which have to be cut for all car types, and which must later be closed up at high cost, if conventional manufacturing techniques are applied.

2.3 LASERS AND ADVANCED MANUFACTURING SYSTEMS

2.3 (1) Development of recuperator manufacturing techniques. Phase 2

Miller, J. A. (Avco Lycoming Division, USA). Final Tech. Rept. No. AD-A132448/2, 116pp. June 1983. (See also Phase 1, Tech. Rept. No. AD-B034 5582).

Describes the development of an automated, computer-controlled, pulsed CO_2 laser welding facility for assembly of a thin plate gas turbine engine recuperator. A detailed analysis and comparison of specific CO_2 laser systems is given, as is an explanation of the general operation of industrial CO_2 lasers and the problems involved in their design and application. The computer control of laser welding systems is discussed with particular emphasis on the use of high speed moving mirror systems to deflect the laser beam around irregular shaped joints. Two computer/moving mirror systems were evaluated and programs for each developed. One was in ESSI, a European machine tool language, and the other in US computer numerical control language. The program developments work and the problems integration of computer and laser systems are discussed. A detailed cost analysis is given.

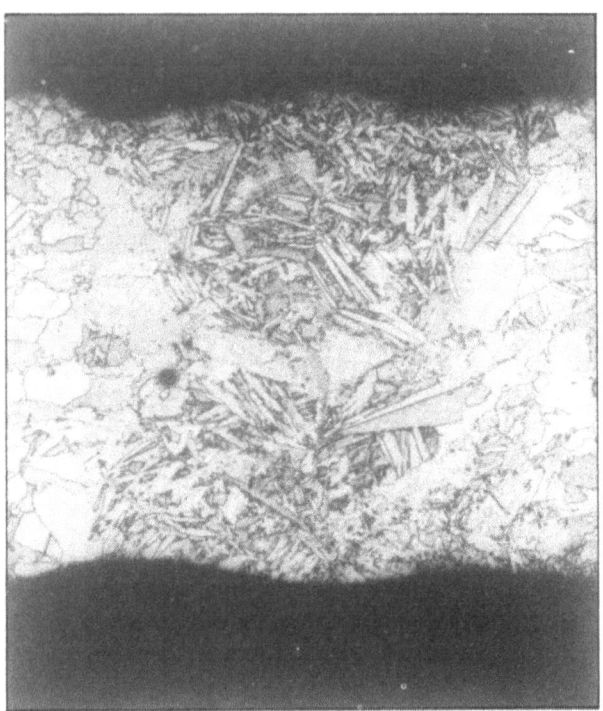

Laser-welded, laser cut edges of commercially pure titanium 1mm thick (See 2.3 (2))

2.3 (2) Development of a multi-workstation laser processing facility

Hardisty, F. B. (Bramah Engineering Ltd, UK). In, *Proc. 1st Int. Conf. on Lasers in Manufacturing*, 1-3 November 1983, Brighton, UK, (Ed. M. F. Kimmitt), pp. 63-70. IFS (Publications) Ltd, Bedford, UK, 1983.

The considerations leading to the purchase of 2kW CO_2 laser and the expansion of the facility from a single workstation to the present system, are outlined. The development of the process for welding and cutting various metals (including steels, nickel-based and titanium alloys) and the way it complements an existing, extensive manufacturing capability, is described. Plans for expansion of the system and further process development (e.g. surface treatments, YAG laser hole drilling) are outlined, as are the potential benefits from combining laser processing with other advanced metalworking processes such as superplastic forming.

2.3 (3) Laser-integrated manufacturing: a proposal to rejuvenate the machine-tool industry

Seaman, F. D. and Rajagopal, S. (IIT Research Institute, USA). *Laser Focus/Electro-Opt.* 19(11): 75-86, 1983.

Proposes a system consisting of a column of laser beams branching out to reach a variety of reconfigurable workstations. The proposal is based on first-hand experience with laser materials processing and on the productivity potential of FMS. It promises an array of manufacturing capabilities in a single system, with cost savings of 10%-30% over conventional batch manufacturing, and more when processing very small batches through several dissimilar operations. (6 refs.)

2.3 (4) A technical and financial appraisal of NC laser machines in a profile cutting application

Janjua, M. S., Rathmill, K. and Allen, D. M. (Cranfield Robotics and Automation Group, UK). In, *Proc. 1st Int. Conf. on Lasers in Manufacturing*, 1-3 November 1983, Brighton, UK, (Ed. M. F. Kimmitt), pp. 41-52. IFS (Publications) Ltd, Bedford, UK, 1983.

At a time when capital available for expenditure on new plant is in short supply, the need to justify investment in improved technology is paramount. This article, which presents a financial and technical case study for the procurement of CNC profiling equipment, is therefore most appropriate. It gives a practical example, comparing the capabilities of different machines together with overall costings and emphasises the importance of matching the choice of machine purchase to specific production requirements.

2.3 (5) CNC lasers cut out saw-blades

Winship, J. T. *American machinist*, 128(5): 92-94, 1984.

A saw-blade manufacturing system is described which combines laser cutting with computerised numerical control (CNC). The laser machines eliminate the need for punching and milling operations and also yield other benefits such as 67% reduction in average production time per blade and substantial savings in tools and dies.

2.3 (6) Computer aided engineering brings quiet revolution to presswork complex

Smith, R. *Met. Construction*, 16(12): 736-738, 1984.

Reports on Austin Rover's Swindon pressworks. Covers developments, including the computer aided design of tools and jigs from a master database definition of an autobody surface, and five axis laser cutting of panels.

2.3 (7) Computer based factory automation — Production research and technology.

Proceedings of the 11th Conference, 21-23 May 1984, Pittsburgh, PA. Society of Manufacturing Engineers, Dearborn, MI, USA, 1984, 420pp.

Comprises 58 papers dealing with various aspects of advanced computer based manufacturing methods and processes. Includes papers on laser processes integrated into AMS.

2.3 (8) Development of a multi-workstation laser processing facility

Hardisty, F. B. (Bramah Engineering Ltd, UK). In, *ICALEO '83, Proc. Materials Processing Symp.*, Vol. 36, 14-17 November 1983, Los Angeles, CA, (Ed. E. A. Metzbower). pp. 174-179. Laser Institute of America, Toledo, OH, USA, 1984.

The considerations leading to the purchase of a 2kW CO_2 laser and the expansion of the facility from a single workstation to the present system are outlined. The development of the process for welding and cutting various metals, (including steels, nickel-based and titanium alloys) and the way it complements an existing, extensive manufacturing capability, is described. Plans for expansion of the system and further process

development, (e.g. surface treatments, YAG laser hole drilling) are outlined, as are the potential benefits from combining laser processing with other advanced metalworking processes such as superplastic forming.

2.3 (9) Development of laminated drawing dies by laser cutting

Kunieda, M. and Nakagawa, T. *Bulletin Jpn. Soc. Precis. Engrs.*, 18(4): 353-354, December 1984.

Illustrates a concept of die-cavity formation by assembling a number of thin sheets with gradually changing profiles cut to suit the cavity geometry. Its application, using a computerised numerically controlled laser cutting machine to obtain the required profiles, is described and assessed.

2.3 (10) Grand FMS almost a runner

Hartley, J. *FMS Mag.*, 2(4): 224-228, 1984.

Reports on the flexible manufacturing system complex at Tsukuba, Japan. The system features laser machining/welding, workpiece handling, assembly cell, and inspection all unmanned. The aims of the complex are the production of assemblies such as gearboxes and spindle housings.

2.3 (11) It's for sheet metal, too — flexible manufacturing systems

Wildish, M. *Mach. and Prod. Engng.*, 142(3): 658, 1984.

The project — supplied as a turnkey system from Trumpf of America — will automate the entire sheet metal stamping operation combining material, tool and data handling. It includes a distributed numerical controlled (DNC), a Trumpf 180 L combined high speed CNC punching and laser cutting machine with fully automatic tool replenishment through a robot.

2.3 (12) Micro controls laser in sheet metal cutting

Janjua, M. S. *Sheet Metal Industries*, 61(10): 558, 561, 562, 564, 1984.

If the number of components to be manufactured in a batch is very small, a microcomputer can store the variables. Since the mathematics of the shapes is presented one can follow this to produce the information required for controlling a laser profile cutter, and the speed of the microcomputer in working out shapes can be

The machining complex of the Tsukuba FMS with the YAG laser room (See 2.3 (10))

made compatible with that of the laser cutting speed. The microcomputer is able to produce control information for the laser cutting system when the point-to-point distance is 20mm and the programming language used is Fortran (4.2.3). The speed can also be increased four times.

2.3 (13) Selecting a CNC control for laser machining systems

Polad, M. In, *Proc. 2nd Biennial Int. Machine Tool Technical Conf.*, Vol. 4, 5-13 September 1984, Chicago, IL, pp. 12, 71-82. National Machine Tool Builders Association, McLean, VA, USA, 1984.

Discusses the CNC features which are necessary to optimise the control of a laser machining system. These include: the features necessary for efficient laser control from the CNC; and the features necessary to hold the desired part tolerances, particularly in relation to the high-speed, lightweight motion systems.

2.3 (14) CAD-CAM laser cutting of saw blades

Billhardt, C. F. and Wittkopp, C. *Werkstatt Betr.*, 118(5): 264-266, 1985. (In German)

Laser cutting devices have been selected for the manufacture of special circular saw blades from Ni steels and Ni-Cr-Mo steels. The machine, together with the organisational steps that have been taken with respect to programming, determination of parameters and economic viability are described. Details are also given

about factory changes, maintenance measures, training questions, and the economic results and the experience that has been gained. (1 ref.)

2.3 (15) Computer aided rationalisation of technological production planning in making blanks for structural steel work in VEB

Schwarz, M. (VEB Schwermaschinenbaukombinat 'Ernst Thalmann'), Germany. *Fertigungstech. & Betr.*, 35(6): 332-334, 1985. (In German)

In the steel work department of the plant, the planning work for producing the punched tapes for running the NC flame cutting machines or the laser cutting unit has been fully automated. Thus, a complete CAM system has been developed.

2.3 (16) Design of FMS with laser

Ikeda, M. (Electrotechnical Laboratory, Japan). In, *ICALEO '84, Proc. Materials Processing Symp.*, Vol. 44, 12-15 November 1984, Boston, MA, USA, (Ed. J. Mazumder), pp. 228-231. Laser Institute of America, Toledo, OH, USA, 1985.

Deals with the concept of the future manufacturing system which is a kind of an ideal system based on the flexibile capability of lasers to produce various types of machine components in a small batch. The system was discussed with the object of shortening of process, economisation of man-hours, dealing with variance of work materials and precise machining not from an economical point of view but from a technical one. A multiplicative production form in which processing, inspection or measurement and assembly are closely integrated was proposed. A laser process for stirling engine manufacture as a model product was discussed.

2.3 (17) A flexible laser manufacturing system based on the composition of several laser beams

Fantini, V. et al. (CISE SpA, Italy). In, *Proc. 2nd Int. Conf. on Lasers in Manufacturing*, 26-28 March 1985, Birmingham, UK, (Ed. M. F. Kimmitt), pp. 249-260). IFS (Publications) Ltd, Bedford, UK, 1985.

An original configuration of a flexible laser manufacturing system is described. This system is based on the concept of the sum of several high power CO₂ laser beams in order to obtain a larger amount of laser power in the workstations starting from relatively low power industrial laser sources.

In the system there are several sources which can deliver beams to the workstations either independently or suitably composed. At any moment the total available laser power can be sent to different workstations which can operate simultaneously.

Moreover, in each workstation it is possible to change the technological process performed simply by automatically substituting the beam manipulating optics.

2.3 (18) High power CO₂ lasers for FMS 10kW and 20kW class SAGE lasers

Tabata, N. et al. (Mitsubishi Electric Corp, Japan). In, *ICALEO '84, Proc. Materials Processing Symp.*, Vol. 44, 12-15 November 1984, Boston, MA, (Ed. J. Mazumder), pp. 238-245. Laser Institute of America, Toledo, OH, USA, 1985.

Industrial 10kW and 20kW CO₂ lasers, which are characterised by high reliability and durability as well as high efficiency and beam quality, have been developed.

These lasers are equipped with a new discharge electrode system for laser gas excitation – SAGE (silent discharge assisted glow excitation) – and KC1 windows for output beam extraction. (See previous papers for a description of the Japanese Flexible Manufacturing System complex provided with laser (FMSC with laser) test plant. Ishihara, K. and others pp. 232-237; Ikeda, M. pp. 226-231).

2.3 (19) Improved metal processing with high power CO₂ lasers welding and cutting; (II) Hardening

Ohmine, M, Nakamura, E. et al. (Mitsubishi Electric Corp, Japan). In, *ICALEO '84, Proc. Materials Processing Symp.*, Vol. 44, 12-15 November 1984, Boston, MA, USA, (Ed. J. Mazumder), pp. 253-260 and 261-268. Laser Institute of America, Toledo, OH, USA, 1985.

A small but complex plate being ejected from the laser cutter (See 2.3 (21))

Report of work at the Flexible Manufacturing System Complex provided with Laser (FMSC) proposing automatically interchangeable laser processing heads, and reporting on the efficiency of welding, cutting and hardening.

2.3 (20) Laser-cutting and welding in a flexible manufacturing system

Uetz, H., Hardock, G. and Warnecke, H.-J. (Fraunhofer-Institute for Manufacturing Engineering and Automation (IPA), West Germany). In, *Proc. 2nd Int. Conf. on Lasers in Manufacturing*, 26-28 March 1985, Birmingham, UK, (Ed. M. F. Kimmitt), pp. 261-278. IFS (Publications) Ltd, Bedford, UK, 1985.

Describes a uniquely flexible manufacturing system planned for a broad spectrum of 3-dimensional sheet metal parts. The 3-dimensional parts are produced from 1.25mm thick sheet metal by stamping, cutting, welding and finishing. Parts assembly, painting and other operations follow. Medium sized series are to be produced in 3 shifts with the third shift being unattended.

2.3 (21) Moving towards FMS for sheet metal

Hartley, J. *FMS Mag.*, 3(4): 207-210, 1985.

Describes the sheet metal FMS installed at Murata Machinery's plant which has produced considerable savings. Describes the use of the AGV, and the production line, concentrating on the separation of partially made products from offcuts. At the laser cutting stage there is no manual material handling, but bending and welding still require a lot of manual work.

2.3 (22) A flexible manufacturing line built around a laser machine

Dullin, E. (CN Industries SA, France). In, *Proc. 3rd Int. Conf. on Lasers in Manufacturing*, 3-5 June 1986, Paris, France, (Ed. A. Quenzer), pp. 365-382. IFS (Publications) Ltd, Bedford, UK, 1986.

Sietam, universally known for its handling installations, has decided to design, realise and use a flexible machining system to modernise one of its plants at Douai and render it profitable. Using advanced technology, this system will manufacture idlers especially designed for coal mines. To answer the needs of the market, there will be 1600 different models of manufactured rollers of all diameters and lengths. The production will be of 400,000 rollers per year, i.e. one roller every 25 seconds. The system will permit a 30% reduction of stocks and 5% of rejects, and will reduce the waste of materials and improve quality, etc.

3

LASER BEAM PHYSICS

3.1 GENERAL

3.1 (1) Induced evaporation of metal from an aluminium surface by a normal pulse neodymium laser

Johnson, C. B. (Naval Postgraduate School Monterey, USA). Rept. No. AD-A08160117; Master Thesis, 77pp., September 1979.

The experiment was conducted using a neodymium glass laser modified for normal pulse operation. The energy density was varied from 850J/sq cm where no breakdown occurred to 5000J/sq cm where the threshold for breakdown was exceeded. The normal pulse duration was 600µs. Analysis of the ejected material was achieved by using a Hughes ionisation gauge placed in the path of the ejected material. Oscilloscope traces of the ionisation gauge output show that the gauge 'sees' what is flying past it. There is good correlation between laser radiation, plasma radiation and ionisation gauge fluctuations. The ionisation gauge gave distinguishable signals for ions, electrons, and neutral particles ejected from the target surface.

3.1 (2) Lasers in the domains of ultrashort time phenomena physics and ballistics

Hugenschmidt, M. (Institut Franco-Allemand de Recherches, France) Rept. No. ISL-CO-236183, 47 pp. October 1983. (In German)

Properties of laser radiation are explained. Laser applications using the high specific energy densities, the focusability of the radiation, and the possibility of energy transport over large distances are mentioned, e.g., in materials machining. The physical mechanisms of the interaction between radiation and matter are explained. Laser measurement instruments such as laser range finders, interferometry, holography, laser scaterring methods, and laser infrared methods are presented.

3.1 (3) Some aspects of the fluid dynamics of laser welding

Dowden, J., Davis, M. and Kapadia, P. (University of Essex, UK). *J. Fluid Mech.*, 126(1): 123-146, 1983.

The equations for the laser welding process are set out and the conditions at the two boundaries reconstructed. Approximate solutions of the problem of low welding speeds are obtained for four different models, and an expression for the minimum power of the laser is calculated.

3.1 (4) Surface active element effects on the shape of GTA, laser and electron beam welds

Aden, R. J. et al. *Weld. J.*, 62(3): 72s-77s, 1983.

Laser and electron beam welds were passed across Se-doped zones in 21-6-9 stainless steel. The depth/width (d/w) ratio of a defocused laser weld with a weld pool shape similar to a GTA weld increased by >200% in a zone where 77ppm Se had been added. Smaller increases were observed in Se-doped zones for a moderately defocused electron beam weld with a higher d/w ratio in undoped base metal. When laser or electron beam weld penetration was by a keyhole mechanism, no change in d/w ratio occurred in Se-doped zones. The results confirm the surface tension driven fluid flow model for the effect of minor elements on GTA weld pool shape. (20 refs.)

3.1 (5) Dynamics of the electro-laser breakdown of metallic materials

Dyatel, V. P. and Kovalenko, V. S. *Elektronnaya Obrab. Mater.*, (3): 15-18, 1984. (In Russian)

Effective hole piercing conditions are set up when localised high pressure pulses are delivered to the discharge channel. This requires close contact, without any gap, between the electrode surfaces in the zone where the electrical discharge releases its energy.

3.1 (6) A new preionization technique for high repetition rate pulsed gaseous lasers

Marchetti, R. B., Penco, E. (Selenia SpA, Italy) and Salvetti, G. (ENEA – CRE Casaccia, Italy). In, *ICALEO '83, Proc. Materials Processing Symp.*, Vol. 38, 14-17 November 1983, Los Angeles, CA, (Ed. E. A. Metzbower), pp. 10-15. Laser Institute of America, Toledo, OH, USA, 1984.

A new preionisation technique for high repetition rate pulsed gaseous lasers to be used in a number of scientific and industrial applications is presented. Performance of a 1kHz CO_2 laser which employs this technique is reported, and the importance of this new method for the realisation of durable, compact and inexpensive devices is discussed. (8 refs.)

3.1 (7) Comparison of the physics of gas tungsten arc welding (GTAW), electron beam welding (EBW), and laser beam welding (LBW)

Nunes, A. C. (National Aeronautics and Space Administration, USA). Rept. No. NAS 1. 15: 86503, 22pp., August 1985. National Aeronautics and Space Administration, Huntsville, AL, USA.

The physics governing the applicability and limitations of gas tungsten arc, electron beam, and laser beam welding are compared. Includes an appendix on the selection of laser welding systems.

3.1 (8) Diagnostics of the process of formation of the welded joint in laser welding using a flat double probe

Gladkov, E. A. et al. (NE Bauman Higher Technical School, USSR). *Weld. Prod.*, 32(3): 51-53, 1985.

The diagnostics are carried out on the basis of the amplitude-frequency characteristics of the probe circuit current.

3.1 (9) Laser processing of high-tech materials at high irradiance

Whitlock, R. R. (US Naval Research Laboratory, USA). In, *Laser Welding, Machining and Materials Processing – Proc. Int. Conf. on Applications of Laser and Electro-Optics, ICALEO '85*, 11-14 November 1985, San Francisco, CA, (Ed. C. Albright), pp. 187-199. Laser Institute of America, Toledo, OH, USA/IFS (Publications) Ltd, Bedford, UK, 1986.

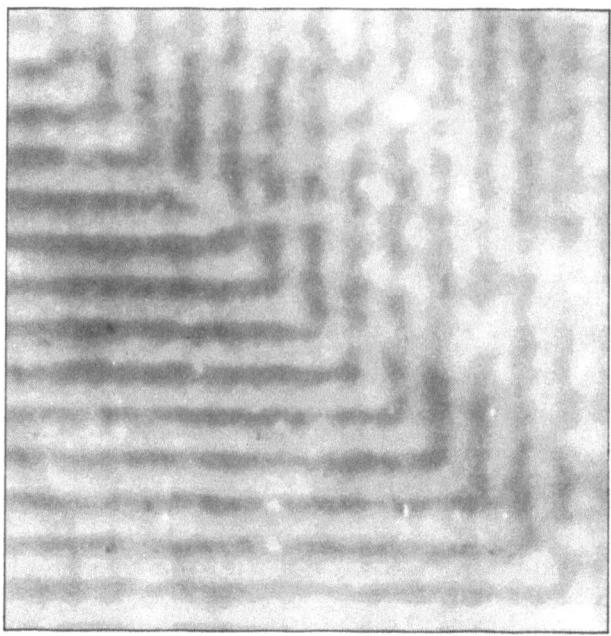

Submicron replication achieved by X-ray lithography with Al plasmas (See 3.1 (9))

The major applications of lasers to the processing of high technology materials are briefly enumerated. It is found that the majority of these applications rely upon the direct thermal effects of irradiating the material with laser light. Several other important interactions of high irradiance lasers with materials are examined. The need for survey and evaluation of the capabilities of lasers to generate shocks for materials processing is underscored. Emphasis is placed on applications and potential applications of secondary processing, in which the particle or photon emissions of a laser-produced plasma are employed in the materials processing steps. The demonstrated feasibility of using X-ray pulses from laser-produced plasmas for X-ray lithographic fabrication of microelectronics and microstructures is briefly reviewed. The possible application of laser evaporative deposition to the fabrication of novel materials, such as multilayers, superlattices, quantum devices and microstructures, is proposed. (57 refs.)

3.1 (10) Laser processing: the use of artificial intelligence and statistics for generating knowledge

Aberkane, Y. and Dumas, M. (Centre d'Etudes Nucléaires de Saclay (CEN/CEA), France). In, *Proc. 3rd Int. Conf. on Lasers in Manufacturing*, 3-5 June 1986, Paris, France, (Ed. A. Quenzer), pp. 77-88. IFS (Publications) Ltd, Bedford, UK, 1986.

The knowledge available in material processing by laser is scattered in the literature and often incomplete. As an example, a good quality of weld or cut is difficult to obtain by direct use of the data and formulae given in the literature.

This paper proposes the first step of a method using both artificial intelligence and statistics for generating new knowledge from experiments and from the data given in the literature for cutting different materials by a CO_2 laser.

3.2 LASER OPTICS

3.2 (1) Optical design problems in industrial solid state lasers

Koechner, W. (Hadron Inc, USA). In, *Proc. SPIE Optical Design Problems in Laser Systems*, Vol. 69, 21-22 August 1975, San Diego, CA, (Ed. W. R. Sooy), pp. 148-157. Society of Photo-Optical Instrumentation Engineers, Palos Verdes Estates, CA, USA, 1975.

3.2 (2) Optics in laser fusion

Trenholme, J. B. (Lawrence Livermore Laboratory, USA). In, *Proc. SPIE Optical Design Problems in Laser Systems*, 21-22 August 1975, San Diego, CA, (Ed. W. R. Sooy), pp. 158-162. Society of Photo-Optical Instrumentation Engineers, Palos Verdes Estates, CA, USA, 1975.

3.2 (3) Linear, annular, and radial focusing with axicons and applications to laser machining

Rioux, M., Tremblay, R. and Belanger, P. A. (Université Laval, Canada). *Appl. Optics*, 17(10): 1532-1536, May 1978.

Most combinations allow continuous adjustment of exit beam parameters, focal line length, focal ring diameter, and magnification, by varying the relative position of one of the axicons. New laser applications are also discussed in relation to these optical devices. (18 refs.)

3.2 (4) The role of focus in heavy-plate laser welding

Seaman, F. D. SME Tech. Paper No. MR78-345, 11pp. 1978. Society of Manufacturing Engineers, Dearborn, MI, USA.

The effectiveness of laser welding processes is shown to be dependent upon beam qualities which include power, mode, freedom from aberrations and geometry. They are brought together in the intense focal region of the beam. Optimising the placement of this region with respect to the work surface has been found to improve process latitude and efficiency and weld quality. Monitoring the behaviour of the focal region in a simple welding test can be used either as a measure of the consistency of laser performance or to transfer procedures from one laser to another.

3.2 (5) Laser welding and drilling

Sharp, C. M. (BOC Industrial Power Beams, UK). In, *Proc. SPIE 4th European Electro-Optics Conf.,* Vol. 164, 10-13 October 1978, Utrecht, Netherlands, pp. 271-278. Society of Photo-optical Instrumentation Engineers, Bellingham, WA, USA, 1979.

Describes the absorption of laser light by a workpiece and shows that the resultant physical, chemical and thermal changes can be controlled to cause either drilling or welding.

3.2 (6) The effects of the focusing conditions on the formation of laser welds

Shovkoplyas, V. M. et al. (E. O. Paton Welding Institute, USSR). *Autom. Weld.*, 33(11): 15-16, 1980.

Relationships for the change in penetration, with the focal plane in various positions, and for lenses with various focal lengths are given.

3.2 (7) Nd: YAG laser beam diagnostics

Pope, L. E. and McDonald, T. G. (Sandia National Laboratories, USA). Rept. No. SAND-81-0011C, CONF-8110110-2, 12pp. October 1981.

Describes a program to develop diagnostic techniques for pulsed Nd: YAG lasers for welding and discusses problems encountered when deviations from ideal optical collimation are incorporated within a welding system. Ideal optical collimation of a laser beam is defined, the diagnostic system is described, and the Sandia welding system is discussed.

3.2 (8) Use of the pulsed-periodic regime of a CO_2 electrical-ionization laser for laser welding

Averin, A. P. et al. (P. N. Lebedev Physics Institute, USSR). *Sov. Phys. Dokl*, 27(10): 4-6, 1982.

Laser welding processes require sharp focusing of radiation and laser regimes with a high off-duty factor are ineffective. So for laser welding, laser regimes with low off-duty factors are of great practical interest, since these regimes make it possible to increase the energy characteristics of the laser and do not decrease the efficiency of the industrial process compared to that in the case of the continuous regime. In addition, the magnitude of the optimum off-duty factor is different for different technological processes and physical properties of the materials being worked.

3.2 (9) Laser beam analyser

Lim, G. C. and Steer, W. M. (Imperial College, UK). In, *Proc. 1st Int. Conf. on Lasers in Manufacturing*, 1-3 November 1983, Brighton, UK, (Ed. M.

F. Kimmitt), pp. 161-168. IFS (Publications) Ltd, Bedford, UK, 1983.

A laser beam analyser is described. It is capable of monitoring the power profile and the spot size at the focus of a high power laser beam. Power fluctuation of the beam can also be monitored, as well as the shape of a pulsed beam. In-process beam monitoring can be carried out with immediate display of the information on an oscilloscope. It is very versatile, convenient to install, easy to operate and has a very high resolution.

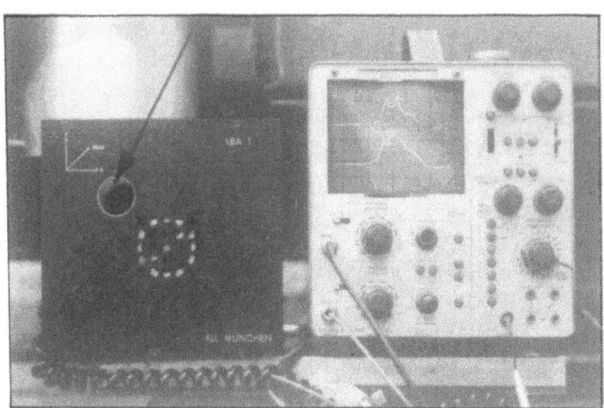

Laser beam analyser and oscilloscope display (See 3.2 (9))

3.2 (10) Selection of optimum beam parameters for laser welding

Skripchenko, A. I. and Surkov, A. V. (Izhorskii Plant Production Association, USSR). *Avtom Svarka*, 359(2): 45-48, 1983. (In Russian)

Analyses the effect of the parameters of focused CO_2 laser radiation on the welding process. Two values characterising the focalised beam are proposed. They permit systematisation of experimental data and formulation of requirements with respect to the optical focusing system to minimise the required laser beam power. (111 refs.)

3.2 (11) Focusing and depth of focus of Gaussian and higher order mode beams

Luxon, J. T. (GMI Engineering & Management Institute, USA). In, *ICALEO '83, Proc., Materials Processing Symp.*, Vol. 38, 14-17 November 1983, Los Angeles, CA, (Ed. E. A. Metzbower), pp. 31-36. Laser Institute of America, Toledo, OH, USA, 1984.

The purpose of this paper is to review and clarify some of the aspects of laser beam propagation, focusing and depth of focus, particularly as related to materials processing lasers. It is hoped that some persistent misconceptions will be dispelled. Simple, correct equations are presented for laser beam propagation, focusing and depth of focus. The well known relationships for Gaussian beams will be reviewed and their applicability discussed. More recent results pertaining to higher-order mode beam behaviour will be presented along with experimental verification of their validity.

3.2 (12) Machining of materials by laser beam. V. Focusing of laser beams, taking the CO_2 laser as an example

Ripper, G. and Herzinger, G. (Technical High School, Darmstadt, Germany). Feinwerktech. & Messtech, 92(6): 297-302, 1984. (In German)

In order to set optimum process parameters for laser cutting, welding and hardening, it is necessary to know the parameters of the focused beam. Of significance are the minimum attainable focal radius which determines the maximum intensity, the depth of focus and the course of the beam caustic behind the focusing lens. (20 refs.)

3.2 (13) Optical fibres for transmitting high-power CO_2 laser beam

Takahashi, K et al. (Sumitomo Electric Industries Ltd, Japan). *Sumitomo Electr. Tech. Review*, 23(1): 203-210, 1984.

There is a demand for development of flexible infrared optical fibre cable to transmit a CO_2 laser beam. Since the CO_2 gas laser emits an infrared beam with a wave length of 10.6µm metal halide crystal such as alkaline halide, silver halide, or thallium halide are the most prospective optical fibre materials. Discusses the fibre forming technique for these crystal materials, and the physical and optical characteristics of the fibres.

3.2 (14) Development of ZnSe optics for high-powered CO_2 lasers

Nanba, H. et al. (Sumitomo Electric Industries, Japan) and Miyata, T. (Matsuschita Research Institute, Japan). In, *ICALEO '84, Proc. Materials Processing Symp.*, Vol. 44, 12-15 November 1984, Boston, MA, (Ed. J. Mazumder), pp. 284-290. Laser Institute of America, Toledo, OH, USA, 1985.

A further paper on KC1 optics by Sakufaji, S. and others follow the paper in the conference proceedings pp. 291-297.

3.2 (15) Diagnosis of CO_2 laser beams

Loosen, P., Beyer, E. and Kramer, R. (Fraunhofer-Institut für Lasertech, Germany). *Laser and Optoelektron*, 17(3): 278-281, 1985. (In German)

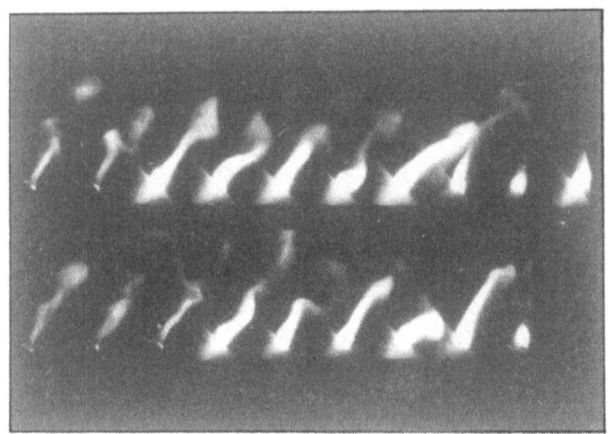

Framing sequence of the interaction zone above the steel target. The sequence is alternating from bottom to top and progresses from left to right (See 3.2 (19))

3.2 (16) Optical fibre sensors in industrial process control

Macfadyen, N. (Barr & Stroud Ltd, UK). In, *Proc. 2nd Int. Conf. on Lasers in Manufacturing*, 26-28 March 1985, Birmingham, UK, (Ed. M. F. Kimmitt), pp. 173-182. IFS (Publications) Ltd, Bedford, UK, 1985.

Reviews the current state of development of optical sensors and identifies those application areas where the power and coherence of a laser make it the preferable, or only possible, light source.

The exploitation of coherence in a sensor requires the use of single mode optical fibre, and, once the information is transmitted in a single mode format, integrated optical circuits are prime contenders for the pre-detector signal processing which is often required. Developments in integrated optics are described, and relevant circuit architectures discussed.

3.2 (17) Optical problems of beam delivery

Forbes, N. (Ferranti Industrial Electronics Ltd, UK). In, *Proc. 2nd Int. Conf. on Lasers in Manufacturing*, 26-28 March 1985, Birmingham, UK, (Ed. M. F. Kimmitt), pp. 309-318. IFS (Publications) Ltd, Bedford, UK, 1985.

Aspects of optical beam delivery are considered. Typical components used in standard beam delivery systems are discussed and material and processing problems associated with the manufacture of such components briefly

mentioned. Modern methods of beam delivery are described, together with projections for future systems. Novel forms of beam delivery are emerging and will be of growing importance to articulated arm and fibre optic delivery systems in future years. The use of mirrors to shape the beam described as well as a production example of a more advanced optical system for cigarette manufacture.

3.2 (18) R & D of optics for high-power CO₂ lasers

Sano, R. et al. (Matsuschita Research Industries, Japan). In, *ICALEO '84, Proc. Materials Processing Symp.*, Vol. 44, 12-15 November 1984, Boston, MA, (Ed. J. Mazumder), pp. 246-252. Laser Institute of America, Toledo, OH, USA, 1985.

Substrate material preparation, surface finishing and coating technologies as well as evaluation technologies are developed to manufacture metal mirrors and transmissive optics made of $ZnSe$ and $KC1$ substrates. These optics are installed to 20kW laser and to CO_2 laser processing subsystem in Tsukuba FMSC Test Plant and are now tested.

(Papers describing the plant are included in the FMS section in this bibliography).

3.2 (19) Optical diagnostics of laser-metal interaction during welding

Schellhorn, M., Nowack, R. and Roth, G. (DFVLR, Institut für Technische Physik, West Germany). In, *Proc. 3rd Int. Conf. on Lasers in Manufacturing*, 3-5 June 1986, Paris, France, (Ed. A. Quenzer), pp.97-106). IFS (Publications) Ltd, Bedford, UK, 1986.

Experimental investigations concerning the basic role of metal vapour and plasma formation during the onset of laser workpiece interaction are reported. The coupling efficiency of CO_2 laser radiation incident at an angle of 45° to the surface of various metallic materials is described. Results of different optical diagnostic techniques are correlated with time dependent absorption measurements of the laser radiation. The plume above the workpiece was analysed spectroscopically to identify the predominant species in the interaction volume. A high speed movie of the interaction zone under different processing conditions is presented.

3.3 LASER THERMO-DYNAMICS AND ENERGY

3.3 (1) Heat balance and flow conditions for electron beam and laser welding

Klemens, P. G. (University of Connecticut, USA). *J. Appl. Physics*, 47(5): 2165-2174, 1976.

Theoretical analyses to determine the factors which govern the shape of the vapour cavity and of the molten zone. Flow conditions in the horizontal plane determine the dimensions of the weld. The balance between beam power and power dissipated by conduction, melting, and vapourisation is discussed and a description is given of cavity formation and beam penetration. (21 refs.)

3.3 (2) Hydrodynamic limit to penetration of a material by a high power beam

Andrews, J. G., Atthey, D. R. and Jerrard, H. G. (Marchwood Engineering Laboratories, CEGB, UK). In, *Proc. Electro-Optics Laser Int. Conf.*, 9-11 March 1976, Brighton, UK, pp. 160-166. IPC Science and Technology Press, Guildford, UK). Also, *J. Appl. Physics D.*, 9(15): 2181-2194, 1976.

In laser welding a high intensity beam is directed on to a metal surface causing melting and evaporation. If the rate of evaporation is sufficiently high the laser will drill a 'keyhole' into the molten metal, thereby depositing power deep into the material. This drilling process will be opposed by the flow of molten metal into the keyhole and in the steady state the two effects balance each other over the entire surface of the hole. Steady state hole profiles are obtained for a vertical beam including the effects of gravity and surface tension. It is shown that surface tension reduces the depth of penetration typically by a factor of three.

3.3 (3) Formation of fusion zone produced by ruby laser

Yoshioka, S., Miyazaki, T. and Miyabe, H. (Chiba Institute of Technology, Japan). *Jpn. Weld. Soc.*, 8(1): 38-41, 1977.

Attempts to explain the formation mechanism of the fusion zone produced on a SUS 304 stainless steel specimen of 0.30mm. thickness during the laser welding process using a single pulse of the ruby laser. Special attention is given to the behaviour of molten metal during the process. The formation of the fusion zone covers the following stages: a piercing hole is formed in the weld immediately after the laser irradiation starts, then molten metal develops, and the growth of the fusion zone begins before the laser irradiation is over (8 refs.)

3.3 (4) Study of infrared color isotherms in controlled welding processes

McGuire, B. C. (Rockwell International, USA). In, *Proc. 3rd Biennial Infrared Inf. Exch.*, 24-26 August 1976, St. Louis, MO, pp. 15-21. AGA Corp, Secaucus, NJ, USA, 1977.

Analytical solutions for predicting the isotherm patterns in a given welding process. Equations are fed into a computer with the end results, a base line as to voltage amperage and travel speed of the torch. Thermovision was used to map the isotherm patterns during the welding process in both gas tungsten Arc Welding and CO_2 laser welding. Colour stills and colour films were used to evaluate the findings and verify the analytical prediction. (2 refs.)

3.3 (5) Heat effect by electron beam and laser welding and heat treatment

Rykalin, N. and Uglov, A. (A. A. Baikov Institute of Metallurgy, USSR). In, *Proc. Int. Centre for Heat and Mass Transfer – Heat and Mass Transfer in Metall. Systems Seminar*, 3-7 September 1979, Dubrovnik, Yugoslavia, 14pp. Hemisphere Publ. Corp, Washington DC, USA.

Principal characteristics of heat sources and the temperature field in materials are analysed. Data are given on physical processes initiated by the action of energy concentrated sources. Structural peculiarities in iron carbonic alloys after laser treatment are considered and problems of liquid phase movement in a thin layer on the hole walls are analysed. Formation of deep penetration is considered.

3.3 (6) Theoretical evaluation of structural technological factors in laser welding of hollow shells

Koledov, L. A. et al. *Svar, Proizvod.*, (5): 4-5, 1980. (In Russian)

Calculations are presented for selecting the type of laser system, to determine the power of a laser impulse and to determine the temperature distribution through the structure. Graphs were built for A1 and brass shells 0.8mm. thick. The temperature of the shell 3mm. distance from the beam impact may reach several hundred degrees, which would damage solder joints situated nearby.

3.3 (7) Comparative study of welding process involving focusing of energy

Robert, H. *Rev. Soudure*, 37(4): 199-210, 1981 and 38(1): 19-25, 1982. (In French)

Study covers laser welding, electron beam welding, and plasma arc welding. The results concerning the characteristics of the heat affected zones obtained by each of these systems are presented. A comparison indicates that if the parts to be welded are thicker than 30mm, the electron beam welding technique appears to be the best choice, while for thicknesses less than 5mm. the laser beam technique should be used.

3.3 (8) Component temperature versus laser welding parameters

Jones, W. H. (Monsanto Res. Corp., USA). Rept. No. MLM-2802, 8pp., February 1981. Monsanto Research Corp. Miamisburg, OH, USA.

Develops a set of laser weld parameters that produce weld joints of high integrity at low temperatures in applications where component temperatures are critically important. Focal distance, pulse rate, welding speed, and beam power are manipulated in varying configurations to produce a leak-tight weld joint at low thermal input.

3.3 (9) Radiative ignition mechanism of solid fuels

Kashiwagi, T. *Fire Safety J.*, 3(3): 185-200, 1981.

Transmittance of external radiation from a CO_2 laser through a boundary layer of decomposition products over a vertical sample surface is measured during the ignition period. Using the experimentally measured incident flux at the sample surface, surface temperature history was calculated from a model that included re-radiation and convection losses from the surface, endothermic decomposition and conduction into the material. The results confirm the significant effect of gas phase absorption on surface temperature (21 refs.)

3.3 (10) Heat source/materials interactions during fusion welding

Eagar, T. W. et al. (Massachusetts Institute of Technology, USA). Rept. No. AD-A118776114, 134pp. April 1982. (See also rept. No. AD-A107 717).

Summarises the second year of research involving spectrographic and thermodynamic studies of vapour emission from welding arcs, infrared measurement of the weld pool surface temperature, fracture toughness of HY-80, sensors for weld automation and laser welding of aluminium.

3.3 (11) Component temperature versus laser welding parameters

Jones, W. H. (Monsanto Research Corp, USA).

Rept. No. MLM-3105 (OP), CONF 8 309158-1, 6pp, September 1983. Monsanto Research Corp, Miamisburg, OH, USA.

To evaluate the thermal excursion experienced at the power cavity wall, a study was conducted using assemblies that had been equipped with 0.05mm. diameter thermocouple wires. The main goal of the study was to determine how changes in the laser welding parameters would affect the powder cavity wall temperature. The study concluded that by utilising a selected set of welding parameters, the design requirement of a 160°C maximum powder cavity wall temperature could be satisfied.

3.3 (12) Effect of gas flow on the penetration depth in CO_2 laser welding of structural materials

Grigor'yants, A. G. et al. (Moscow Higher Technical School, USSR). *Avtom. Svarka*, 369(12): 38-40, 1983. (In Russian)

Analyses of the physical processes taking place in the protective gas. Increasing the penetration depth irrespectively of the gaseous medium employed is proposed by additional flow of gas. The mass gas flow rate is determined from the plasma energy balance, taking account of the parameters of the operating conditions and radiation focusing. (22 refs.)

3.3 (13) Heat and mass transfer in electron beam and laser welding and heat treatment

Rykalin, N. N. and Uglov, A. A. (Institute of Metallurgy, USSR). *Heat Transfer – Sov. Res.*, 15(3): 20-36, 1983.

Presents a survey of heat and mass transfer in electron beam and laser welding and heat treatment including characteristics of heat sources and temperature fields in materials. Data on several processes initiated by concentrated energy sources are listed. Structural modification in carbon iron alloys following laser heat treatment is described. Deep welding of metals by electron and laser beams is covered as are studies of flow of the molten metal in formation of deep welds by concentrated energy sources. (48 refs.)

3.3 (14) Hole formation in tellurium alloy films during optical recording

Blom, G. M. *J. Applied Physics*, 54(11): 6175-6182, 1983.

It is shown that the temperature gradient resulting from a focused laser beam creates minimum surface tension at the hottest point of the molten spot. Consequently, a shear stress pulls material from the centre towards the edge, forming a rim and a hole. The cross-sectional shape of the rim is modelled and experimental data on cross-sections of rims around fully opened holes and around optical dropouts support the model, showing rounded rims for the first and very steep edges surrounding for the latter. (23 refs.)

3.3 (15) An investigation of the laser cutting process with the aid of a plane polarized CO_2 laser beam

Lepore, M. et al. *Optic and Lasers Engineering*, 4(4): 241-251, 1983.

Gives further experimental support to the theory of the existence of two different regimes in the oxygen-assisted cutting of steels and explains the dependence of the process on $Fe-O_2$ exothermic reactions. (8 refs.)

3.3 (16) Thermal cycle in the weld zone in laser welding

Volobuev, Yu. V. (N. E. Bauman Higher Technical School, USSR). *Weld. Prod.*, 30(9): 34-35, 1983.

The heating and cooling rates in the heat affected zone in laser welding steel were obtained for the first time by experiment. (33 refs.)

3.3 (17) Absolutely uniform illumination of laser fusion pellets

Schmitt, A. J. (Naval Research Laboratory, USA). Final rept. No. NRL-MR-5221, 11pp., February 1984. Naval Research Laboratory, Washington, DC. USA).

Absolutely uniform illumination of spherical laser fusion pellets is possible when the energy deposition from a single laser beam is given by a simple cos 3 theta distribution. Conditions can be derived for which the laser beam targeting angles allow this absolute illumination uniformity. Configurations based upon the cube and higher order (Platonic solids satisfy the constraints, as well as an infinite class of other less symmetric configurations.)

3.3 (18) Basic analysis of metal removal neodymium lasers

Jones, M. G. and Georgalas, G. (GEC, USA). In, *ICALEO '83, Proc., Materials Processing Symp.*, Vol. 38, 14-17 November 1983, Los Angeles, CA, (Ed. E. A. Metzbower), pp. 199-207. Laser Institute of America, Toledo, OH, USA, 1984.

The laser drilling process has been modelled using a first-order thermal balance approach. An algorithm defining the coupling of laser pulse energy with metals has been derived. Incorporated in this algorithm are material, optical, and geometrical properties required for adequate system modelling. Tests on both a neodymium-glass and a neodymium: YAG (yttrium-aluminium-garnet) based laser have been conducted. Results with two sets of optical settings indicate an excellent correlation between predicted and actual pulse energies. (5 refs.)

3.3 (19) Beam profile measurement of high power CO_2 laser

Ikeda, M., Yamada, A. and Shinohara, K. (Electrotechnical Laboratory, Japan). In, *ICALEO '83, Proc. Materials Processing Symp.*, Vol. 38, 14-17 November 1983, Los Angeles, CA, (Ed. E. A. Metzbower), pp. 16-21. Laser Institute of America, Toledo, OH, USA, 1984.

Power density distribution, called beam profile, of high power CO_2 laser beam is successfully measured with the system designed especially. Incident beam is damped to match with the sensitivity of a detector using a new beam splitter. Data measured at 200×200 points in cross section of a beam are stored into IC memories. Power density distribution, contour lines, a bird's-eye view etc. are calculated with microprocessors and displayed on a CRT.

3.3 (20) Estimating effects of processing conditions and variable properties upon pool shape, cooling rates, and absorption coefficient in laser welding

Chande, T. and Mazumder, J. (University of Illinois, USA). *J. Applied Phys.*, 56(7): 1981-1986, October 1984.

An examination is made of the role of traverse speed, Beer-Lambert absorption coefficient, surface reflectivity, and changing liquid thermal conductivity upon the shape of the melt pool and the cooling rates that occur. With an increase in traverse speed, the pool flattens out and is swept back, and cooling rates increase. (122 refs.)

3.3 (21) Heat transfer, fluid flow and element vaporization in laser welding

DebRoy, T. (Pennsylvania State University USA). Progress Rept. No. DOE/ER/45030-1 25pp, March 1984.

Iron manganese and chromium were the most dominant species in the plasma. A novel technique for the determination of weld pool temperature is presented. It is demonstrated that the relative rates of vapourisation of any two elements from the molten pool can serve as an indicator of the weld pool temperature, irrespective of the element pair selected.

3.3 (22) Calculated transient two-dimensional marangoni flow in a pulsed laser weld pool

Russo, A. J. (Sandia National Laboratories, USA). Rept. No. SAND-85-0225C, CONF-851125-1, 23pp., April 1985.

Pulsed laser welding gives weld pools which have strong radial temperature gradients at the pool surface. The flow in the molten pool is of interest because it can change the temperature distribution in and around the molten zone altering the weld shape, and because it controls the transport of surface contaminants or alloys into the substrate. A two-dimensional finite difference code is being developed to model the transient flow in a prescribed axisymmetric weld pool geometry. A flux corrected transport algorithm is used with a vorticity-steam function formulation of the problem to calculate the flow field. Flow velocities during a typical laser pulse are presented for nickel and steel.

3.3 (23) The dimensions of joints produced by laser beam in metal laser beam welding

Kostrubiec, F. (Lodz Technical University, Poland). *Zesz. Nauk. Politech. Lodz. Elektr.*, (77): 63-74, 1985. (In Polish)

Describes the method of calculation of laser beam energy parameters which secures the production of a joint. Results of laboratory experiments are also given. (1 ref.)

3.3 (24) Energy balance in gas assisted laser-metal interaction

Donati, V. (CISE SpA, Segrate, Italy). In, *Proc. 5th Int. Symp. on Gas Flow and Chemical Lasers*, 20-24 August 1984, Oxford, UK, (Eds. A. S. Kaya and A. C. Walker). Adam Hilger, Bristol, UK, 1985, 534pp.

An experimental research programme is being carried out on the phenomena related to laser-metal interaction, specifically under conditions met in laser welding. Results from backscattered 10.6nm. radiation measurements and calorimetric evaluation of the absorbed energy are used for an energy balance.

3.3 (25) Experimental and theoretical studies on transport processes in laser welding

DebRoy, T. (Pennsylvania State University, USA). Rept. No. DOE/ER/45158-2, 44pp., October 1985.

If an alloy contains one or more volatile components, selective vapourisation of relatively more volatile components during laser welding may lead to inadequate control of the weld composition and poor mechanical properties of the fabricated product. The loss of alloying elements and the eventual properties of the weld zone are influenced by fluid flow in the weld pool, heat transfer and the thermodynamics and the kinetics of the vapourisation of various components from the molten pool. In order to understand and eventually control the alloying element loss from the laser melted pool an experimental and theoretical programme of studies was begun.

3.3 (26) High temperature vapor pressures of metals from laser evaporation

Bober, M. and Singer, J. (Kernforschungszentrum Karlsruhe, Institut für Neutronenphy. und Reaktortech, Germany). *High Temp Sci.* (4th Int. Conf. on High Temperature and Energy Related Materials, 1-6 April 1984, Santa Fe, NM) 19(3): 329-345, 1985.

A newly developed non-stationary technique is applied to measure the high-temperature vapour pressures of metals from laser vapourisation. A fast pyrometer, and ion current probe, and an image converter camera are used to detect incipient boiling from the time-temperature curve. The saturated-vapour pressure curves of stainless steel (Type 1.4970) cladding, and of molybdenum are experimentally determined in the temperature ranges of 2800-3900 and 4500-5200K, respectively. (25 refs.)

3.3 (27) Intensity profile measurement of focused CO_2 laser beam using PMMA

Miyamoto, I., Maruo, H. and Arata, Y. (Osaka University, Japan). In, *ICALEO '84, Proc. Materials Processing Symp.*, Vol. 44, 12-15 November 1984, Boston, MA, USA, (Ed. J. Mazumder), pp. 313-320. Laser Institute of America, Toledo, OH, USA, 1985.

A method by which intensity profile and diameter of focused CO_2 laser beam are determined from the evaporated shape in PMMA (polymethyl methacrylate) has been developed. This method utilizes the fact that PMMA is sublimated by intense laser beams with negligible thermal conduction and reflection losses. Two essential thermal constants of PMMA for estimating the intensity of the laser beam are determined; energy for evaporation H=3000J/cm^3, and threshold energy for evaporation G=6J/cm^2. It is shown that thus estimated profile is in good agreement with the profile determined by the probe techniques, and that the axi-symmetrical intensity distribution is determined from the deconvolution by Harker method. The optimum scanning speed, and tolerable scanning speed range in this technique are given as a function of power and spot size of laser beam. It is also shown that arbitrary intensity distributions can be determined by using PMMA. (21 refs.)

3.3 (28) Molten region temperature distribution in laser welding

Dowden, J. (Essex University, UK). *J. Physics, D*, 18(10): 1987-1994, 1985.

When penetration welding is performed with a laser, the boundary of the keyhole formed in the weld pool near to the surface of the workpiece experiences considerable variation of temperature; this variation could affect the dynamics of the flow in a number of ways. A simple mathematical model for the temperature distribution is investigated, and solutions obtained. These confirm that for values of the Peclet number of order one the variation of temperature at the keyhole surface can be very substantial. (11 refs.)

3.3 (29) Temperature profiles induced by a stationary CW laser beam in a multi-layer structure – application to solar cell interconnect welding

Oh, J. E., Ianno, N. J. and Ahmed, A. U. (Nebraska University, USA). *Appl. Phys. Commun*, 5(3): 113-138, 1985.

Takes into account the temperature dependence of the thermal conductivity and diffusivity as well as free carrier absorption of the incident beam in the silicon where appropriate. Finally, the theoretical temperature profiles are used to determine the weld spot size and these values are compared with results obtained from a simple welding experiement. (18 refs.)

3.3 (30) Thermal field modelling, energy recovery and material addition methods for laser processing

La Roca, A. V. (Fiat Auto, Italy). In, *ICALEO '84, Proc. Materials Processing Symp.*, Vol. 44, 12-15 November 1984, Boston MA, (Ed. J. Mazumder), pp. 198-225. Laser Institute of America, Toledo, OH, USA, 1985.

3.3 (31) CO$_2$ laser design considerations for pulsed material processing

Sasnett, M. W. (Coherent General Inc. USA). In, *Proc. 3rd. Int. Conf. on Lasers in Manufacturing*, 3-5 June 1986, Paris, France, (Ed. A. Quenzer), pp. 279-292. IFS (Publications) Ltd, Bedford, UK, 1986.

Scaling relationships for both continuous (CW) and enhanced pulse performance of slow flow and fast flow CO$_2$ lasers are derived using heat flow analysis. It is shown that CW power for these two laser types scales with active length and mass flow rate, respectively. Under conditions of pulsed electrical excitation of short duration, however, the maximum pulse energy for both types scales with the total heat capacity of the gas in the active region. Materials processing consequences resulting from the different relationship between CW and pulsed performance for the two types is discussed.

3.3 (32) High speed photographic study of YAG laser materials processing

Matsunawa, A. and Katayama, S. (Welding Research Institute of Osaka University, Japan). In, *Laser Welding, Machining and Materials Processing – Proc. Int. Conf. on Application of Lasers and Electro-Optics, ICALEO '85*, 11-14 November 1985, San Francisco, CA, (Ed. C. Albright), pp. 41-48. Laser Institute of America, Toledo, OH, USA/IFS (Publications) Ltd, Bedford, UK, 1986.

The fluid mechanical structure and properties of laser induced plume were systematically clarified by the high time resolution shadowgraph and Schlieren photographic methods using a 40ns pulsed ruby laser as a light source. Also observed was the temporal change of fusion nugget of target surface melted by YAG laser irradiation and the relation between plume generation and molten liquid behaviour was revealed.

3.3 (33) Interest of isothermal power-velocity-radius (P, v, r) diagrams in continuous laser beam characterisation

Dietz, J. and Merlin, J. (Institut National des Sciences Appliquées (INSA), France). In, *Proc. 3rd Int. Conf on Lasers in Manufacturing*, 3-5 June 1986, Paris, France, (Ed. A. Quenzer), pp. 359-364. IFS (Publications) Ltd, Bedford, UK, 1986.

3.3 (34) Measurement of the temperature distribution in CW laser heated materials

Jeanloz, R. and Heinz, D. L. (University of California, USA). In, *Laser Welding, Machining and Materials Processing – Proc. Int. Conf. on Applications of Lasers and Electro-Optics, ICALEO '85*, 11-14 November 1985, Laser Institute of America, Toledo, OH, USA, San Francisco, CA, pp. 239-243. IFS (Publications) Ltd, Bedford, UK, 1986.

Temperatures between about 1500 and 7000K achieved in samples under CW irradiation from a Nd:YAG laser are successfully measured with a spectroradiometer operating at wavelengths of 400 to 850nm. Absolute accuracy is 200K or better, as confirmed by observing the melting points of metals between 1800 and 3700K. Fluctuations in laser output have been reduced to 1-3% (rms), but these still produce temperature fluctuations of 10 to 15% (~200 to 1000K). Two-and-three-dimensional temperature distributions are obtained by a slit-scanning (tomographic) technique. The change from low values of peak temperature (~1500K) and temperature gradients (~10K/μm) at low laser power (~7W) to large values (~5500K and ~400K/μm) at high laser power (~20W) is illustrated for a Mg-silicate ceramic of geophysical importance.

3.4 PLASMA PHENOMENA AND PLUMES

3.4 (1) Plasma phenomena during Nd: YAG laser welding

Casey, H. (Los Alamos Sci. Lab, NM, USA). In, *IEE Conf. Publ.*, 189, part 2, *Proc. 6th Int. Conf. on Gas Discharges and Their Applications*, 8-11 September 1980, Edinburgh, UK, pp. 86-89. IEE, London, UK, 1980. (Summary of Los Alamos Sci. Lab. Rept. No. LA-UR-80-10688 CONF 800922-2).

The laser is used to perform a seal weld at pressure and thus produce a gas-filled sealed IF target. This is a sensitive welding procedure. Difficulties in obtaining consistent welding results, particularly as the gas pressure is increased above approximately 7MPa are reported. (3 refs.)

3.4 (2) Pulsed CO_2 laser welding

Albright, C. E. In, *Proc. Trends in Welding Research in the United States Conf.*, 16-18 November 1981, New Orleans, LA, pp. 653-665. American Society for Metals, OH, USA, 1982.

Reflectivity and plasma plume effects limit the laser light heat input into metal surfaces. Pulsed CO_2 laser welding has an initial power spike which aids in overcoming the reflectivity barrier. Heat input efficiencies $> 30\%$ have been measured using helium to suppress plasma plume formation. Plasma plumes in Ar lift away and decouple the input energy from the metal surface and heat input efficiencies are limited to 15%. Pulsed CO_2 laser welding can be used to break up the centreline plane and prevent centreline solidification cracking in high-speed laser welding of 316 and 4130 steels.

3.4 (3) CO_2 laser welding with plasma utilisation

Minamida, K. et al. In, *ICALEO '82, Proc. Materials Processing Symp.*, Vol. 31, 20-23 September 1982, Boston, MA, pp. 65-72. Laser Institute of America, Toledo, OH, USA, 1982.

The new laser welding method increases welding efficiency and improves the quality and shape of welded joints. Plasma, generated by the interaction between the laser beam, the metal vapour and the laser assist gas, is pushed into the keyhole with a plasma control gas and utilised effectively rather than removed as in the conventional methods. This method is suitable for the coil build-up welding in continuous coil processing in the steel industry and an example of its use in welding 304SS is given.

3.4 (4) Effect of assist gas on bead formation in high power laser welding

Arata, Y., Oda, T. and Nishio, R. (Osaka University, Japan). *Jpn, Weld, Res. Inst.*, 12(2): 161-166, 1983.

Studies role of assist gas in deep penetration welding with a high power CO_2 laser. It is found that there exists a transition pressure above and below which three different types of bead shape are produced. When the assist gas pressure is equal to or a little higher than that of the plasma, the penetration depth increases, and over this pressure the beam hole begins to be expanded from the surface reducing the wall-focusing effect.

3.4 (5) Beam hole behaviour during laser beam welding

Arata, Y., Abe, N. and Oda, T. (Welding Research Institute, Osaka University, Japan). In, *ICALEO '83, Proc. Materials Processing Symp.*, Vol. 38, 14-17 November 1983, Los Angeles, CA, (Ed. E. A. Metzbower), pp. 59-66. Laser Institute of America, Toledo, OH, USA, 1984.

Fundamental phenomena during laser beam welding in steel and glass, including beam hole shape, molten metal flow and peculiar plasma behaviour, were observed dynamically using a transmission X-ray system and high speed camera. The effect of altering the flow rate of helium assist gas was also studied. It was found that the gas flow rate had a strong effect on the beam hole shape, molten metal flow and plasma production. In order to avoid the interference effect of laser plasma, a new laser welding process, named "Laser Spike Seam Welding", was developed. This process allows considerably deeper penetration than conventional continuous welding. The reasons for this superiority were analysed by the above high-speed film method. (12 refs.)

3.4 (6) Development and optical absorption properties of a laser induced plasma during CO_2 laser processing

Bever, E. et al. (Tech. Hochschule, Darmstadt, Germany). In, *Proc. SPIE Int. Soc. Optic. Engineers*, Vol. 455, 26-27 September 1983, Linz, Austria, pp. 75-80. SPIE, Austrian Phys. Soc., 1984.

Laser material processing is accompanied by a laser induced plasma in front of the target surface as soon as the laser radiation exceeds a certain critical intensity. For CW CO_2 laser machining of metal targets the threshold for plasma onset is about $106W/cm^2$. Critical condition for plasma generation at this intensity level is to reach evaporation temperature at the target's surface. At intensity levels exceeding $106W/cm^2$ the laser light is interacting with the laser induced plasma and then the plasma in turn interacts with the target. The absorptivity is no longer constant, but increases with increasing intensity of the incident radiation, so that the total amount of power coupled to the target is increasing. This holds up to intensity levels of $2.107W/cm^2$. Then the plasma begins to withdraw from the target surface, thus interrupting plasma-target interaction.

3.4 (7) A statistical approach to non-destructive testing of laser welds

Duncan, H. A. *J. Eng. Mater. Technol.* 105(3):224-229, 1983

Presents a statistical analysis of the data obtained from a relatively new non-destructive technique for laser welding. Information relating to the quality of the welded joint is extracted from the high-intensity plume which is generated from the materials that are being welded, giving a numerical value associated with the material vapourisation and, consequently, the weld quality. Optimum thresholds for the region in which a weld can be considered as acceptable are determined based on the Neyman-Pearson criterion and Bayes rule. (11 refs.)

3.4 (8) The influence of a plasma during laser welding

Dixon, R. D. and Lewis, G. K. (Los Alamos National Laboratory, USA). In, *ICALEO '83, Proc. Materials Processing Symp.,* Vol. 38, 14-17 November 1983, Los Angeles, CA, (Ed. E. A. Metzbower), pp. 44-50. Laser Institute of America, Toledo, OH, USA, 1984.

High-speed films of single pulse laser welds are correlated with optical emission from the plasma and the acoustic wave to study and model enhanced coupling. The films and acoustic signals combine to support enhanced coupling through the development of a laser supported combustion wave.

3.4 (9) Laser plasma X-ray source for X-ray lithography

Epstein, H. M. (Battelle's Columbus Laboratories, USA). In, *ICALEO '83, Proc. Materials Processing Symp.,* Vol. 38, 14-17 November 1983, Los Angeles, CA, (Ed. E. A. Metzbower), pp. 1-9. Laser Institute of America, Toledo, OH, USA, 1984.

Laser plasma X-ray sources have been evaluated for submicron X-ray lithography. Exposure machines based on available, repetitively pulsed lasers of reasonable cost appear to be attractive. These machines would make full wafer exposure levels in times consistent with present manufacturing requirements. (18 refs.)

3.4 (10) Plasma plume effects in pulsed carbon dioxide laser spot welding

Arnot, R. S. and Albright, C. E. (Ohio State University, USA). In, *ICALEO '83, Proc. Materials Processing, Symp.,* 14-17 November 1983, Los Angeles, CA, (Ed. E. A. Metzbower), pp. 51-58. Laser Institute of America, Toledo, OH, USA, 1984.

The effect of pulsed CO_2 laser spot welding variables on plasma plume formation in argon and helium, and resulting spot weld characteristics in AISI 1015 steel have been investigated.

Two penetration modes were found to occur. At high peak power densities, and with the focal point held at the target surface, a vapour cavity penetration mode was observed. This was associated with a blue plasma plume which absorbed laser energy. At low peak power densities, or with the focal point above the target surface, a heat conduction penetration mode was observed. This mode was associated with a red vapour plume which did not significantly absorb laser energy. Both plume types were larger in argon than in helium. In addition, in argon at higher power densities, the blue plasma plume associated with the vapour cavity penetration mode can decouple from the target surface preventing further heat input. (7 refs.)

3.4 (11) Porosity decrease in laser welds of stainless steel using plasma control

Estill, W. B. and Formisaro, B. D. (Sandia National Laboratories, USA). In, *ICALEO '83, Proc. Materials Processing Symp.,* Vol. 38, 14-17 November 1983, Los Angeles, CA, (Ed. E. A. Metzbower), pp. 67-72. Laser Institute of America, Toledo, OH, USA, 1984.

High energy laser welding incorporating plasma control has been studied and reported by numerous investigators. These investigators demonstrated significant increases in laser weld penetration by use of plasma control.

In this report we show, in addition to variations in weld penetration, drastic decrease in porosity and variation in weld bead shapes resulting from laser welds incorporating plasma control. In particular, deep laser welds (greater than 6mm) have been produced in 304L stainless steel that show no root porosity and only very few, if any, detectable micropores. (22 refs.)

3.4 (12) A simple method for detecting the time profile of a CO_2 laser pulse using a methyl-methacrylate plate

Dell'Erba, M. et al. (Centro Laser, Italy). *Appl. Phys. Commum,* 4(1): 49-61, 1984.

3.4 (13) Beam-plume interaction in pulsed YAG laser processing

Matsunawa, A. et al. (Welding Research Institute, Osaka University Japan). In, *ICALEO '84, Proc. Materials Processing Symp.,* Vol. 44, 12-15 November 1984, Boston, MA, (Ed. J. Mazumder), pp. 35-42. Laser Institute of America, Toledo, OH, USA, 1985.

Experimental studies on beam-plume interaction were conducted when a pulsed Nd:YAG laser was irradiated on a titanium target in the pressure range of 10Pa to $3 \times 10^5 Pa$. In this paper are described the energy dissipation processes of incident beam, constituent and fluidmechanical structure of plume, absorption and scattering of incident radiation in plume, and their effects on energy transfer of laser beam to target material.

3.4 (14) Control of magnesium loss during laser welding of A1-5083 using a plasma suppression technique

Blake, A. and Mazumder, J. (University of Illinois, USA). *J. Engineering Ind.,* 107(3): 275-280, 1985. This is an update of a paper in, *ICALEO '82, Materials Processing Symp.,* Vol. 31, 20-23 September 1982, pp. 33-50. Laser Institute of America, Toledo, OH, USA, 1982.

Loss of magnesium results in welds of low tensile strength and unacceptable porosity. A method has been developed to control plasma formation during welding, resulting in satisfactory welds with little or no magnesium loss and porosity. Plasma formation was controlled by manipulation of a main gas jet in a shielding arrangement during welding. The experiment was carried out using a 10kW CW-CO_2 laser; a two-level factorial experimental design correlated the effects of the independent laser processing variables with as-welded alloy chemistry.

3.4 (15) Diagnostics of the process of weld seam formation during laser welding by means of a flat double probe

Gladkov, E. A. et al. *Svar. Proizvod.,* (3): 40-42, 1985. (In Russian).

A flat electric double probe was effectively used for monitoring the temperature of a plasma flame during laser welding without a protective gas. When gas was fed into the weld zone, the change in current amplitude and frequency in the probe circuit was used to monitor the weld seam formation process. An automatic laser welding control system using amplitude-frequency characteristics of a double probe is described. (3 refs.)

3.4 (16) Effect of angle of incidence on plasmas generated during laser welding

Dixon, R. D. and Lewis, G. K. (Los Alamos National Laboratory, USA). In, *ICALEO '84, Proc. Materials Processing Symp.,* Vol. 44, 12-15 November 1984, Boston, MA, (Ed. J. Mazumder), pp. 28-34. Laser Institute of America, Toledo, OH, USA, 1985. (See also rept. No. LA-UR-84-3808 CONF-8411100-4, 8pp. 1984).

Extension of previous work attempting to verify the existence of laser supported combustion wave, and to provide additional measurements of the plasma initiation time. (6 refs.) (See Welding Journal and ICALEO '83 for earlier work.)

3.4 (17) Fundamental phenomena in laser welding

Arata, Y., Abe, N. and Oda, T. (Osaka University, Japan). In, *Gas Flow and Chemical Lasers, Proc. 5th Int. Symp.,* 20-24 August 1984, Oxford, UK, (Eds. A. S. Kaye and A. C. Walker), pp. 61-66. Adam Hilger, Bristol, UK, 1985.

Fundamental phenomena including beam hole shape and plasma behaviour were observed dynamically using a transmission X-ray system and a conventional high speed carnera. These phenomena were analysed and a new welding process 'laser spike seam welding' was proposed. The dynamic behaviour of the beam hole and plasma under vacuum condition was also observed.

3.4 (18) Laser beam interactions with vapor plumes during Nd:YAG laser welding on aluminum

Peebles, H. C. et al. (Sandia National Laboratories, USA). Rept. No. SAND-84-2667C CONF-8504128-2-Abst., 2pp., 1985.

3.4 (19) Laser produced plasma effects on welding

Dixon, R. D. and Lewis, G. K. (Los Alamos National Laboratory, USA). Rept. No. LA-UR-85-994 CONF-850979-1, 4pp, 1985.

Acoustic data and high speed pictures indicate that a laser-supported combustion wave is generated and propagates away from the sur-

face in the direction toward the laser. The high speed pictures also show that numerous plasmas are generated during one laser pulse. The depth of penetration is also dependent upon the number of plasmas generated, with more plasmas producing more penetration. This latter fact supports the conclusion that the plasma improves the efficiency of the laser process.

3.4 (20) Nd:YAG laser welding experiments

Akau, R. L. et al. (Sandia National Laboratories, USA). Rept. No. SAND-85-1523C CONF-8510170-1, 18pp, October 1985.

Laser beam/plume interaction experiments were conducted using a high speed camera to study plume growth phenomena and to determine maximum plume velocities. Tests were done on four different metals: aluminium 1100, molybdenum, nickel 200, and 304 stainless steel. Previous laser welding experiments have indicated that the vapour plume ejected from the irradiated base material significantly attenuates the laser beam energy for nickel 200 and stainless steel 304. To substantiate this observation, the plume was subjected to a cross flow of argon gas. Metallurgical studies showed a significant increase in weld penetration for all materials except for aluminium. Thus, the specimen was tilted at different angles in an attempt to reduce laser beam attenuation. Results showed no significant increase in weld depth when the tilt angle was increased.

3.4 (21) Plasma heating effects during laser welding

Lewis, G. K. and Dixon, R. D. (Los Alamos National Laboratory, USA). Rept. No. LA-UR-85-1141 CONF-850975-1, 11pp., 1985.

Control of the process depends on an understanding of the laser-plasma-material interaction and characterisation of the laser beam being used. Inherent plasma formation above the material surface and subsequent modulation of the incident laser radiation directly affect the energy transfer to the target material. The temporal and spatial characteristics of the laser beam affect the available power density incident on the target, which is important in achieving repeatability in the process. Other factors such as surface texture, surface contaminants, surface chemistry, and welding environment affect plasma formation which determines the weld penetration.

3.4 (22) The role of assist gas in CO_2 laser welding

Miyamoto, I., Maruo, H. and Arata, Y. (Osaka University, Japan). In, *ICALEO '84, Proc. Materials Processing Symp.*, Vol. 44, 12-15 November 1985, Boston, MA. (Ed. J. Mazumder), pp. 68-75. Laser Institute of America, Toledo, OH, USA, 1985.

Optimum gas-assisting parameters and their tolerable setting errors in controlling plasma in CO_2 laser welding at 1 and 10kW power levels were determined by using a system by which height and angle of the nozzle, assist gas-workpiece interaction position and pressure of the assist gas can be precisely adjusted for various gas species and nozzle diameters. The role of the assist gas in controlling plasma was clarified on the basis of pressure measurements of assist gas and vapour in the cavity, high speed motion pictures and measurement of plasma brightness. It was found that the assist gas suppresses the plasma at pressures slightly higher than the vapour pressure by forcing the vapour to flow away from the focused laser beam along the cavity rear wall, providing deep weld bead without weld defects. A method of monitoring plasma control by using phototransistor is also proposed. (16 refs.)

3.5 LASER SYSTEM MODELLING

3.5 (1) Finite element analysis of laser welding induced thermal shock

Sidorowicz, K. (Bendix Corp, USA). In, *Proc. Conf. on Applications of Lasers in Material Processing*, 18-20 April 1979, Washington, DC, pp. 65-81. ASM (Mater. Metalwork Technol. Ser.) 1979. Also, Bendix. Corp. Rept. No. BDX-613-2118 CONF-78 04104-1, 21pp, 1979; and also in *Welding J.*, 58(11): 3245-3295, 1979.

The thermal and mechanical responses of a girth fillet weld are approximated by the finite element method and an isothermal ring heat source. Approximately 0.1ms after the application of the ring heat source, the radial traction exceeds the failure strength in tension, followed by fracture. Swelling and elongation are in evidence near the ring heat source. The analytical model gives an insight into the physical behaviour of the weldment during CO_2 laser beam welding. (14 refs.)

3.5 (2) Heat transfer model for CW (continuous wave) laser material processing

Mazumder, J. and Steen, W. M. *J. Appl. Phys.*, 51(2): 941-947, 1980.

A three-dimensional heat transfer model for laser material processing with a moving Gaussian heat source is developed using finite difference numerical techniques.
Discusses advantages of this method over others together with comparisons between the model predictions and experiments in laser welding, laser arc welding, laser surface treatment and laser glazing. (42 refs.)

3.5 (3) Two phase mechanism of laser-induced removal of thin absorbing films: I — Theory

Jakovlev, E. B. et al. *J. Phys. D (Appl. Phys.)*, 13(8): 1565-1570, 1980.

The qualitative concepts of melt surface evaporation and of melt motion under the action of the reactive vapour pressure, of surface tension forces and of adhesion, support the model. The basic equations describing the removal process are solved numerically. An analytical parameter phi, characterising the relationship between the liquid and vapour phases of the removed material, is obtained. (13 refs.)

3.5 (4) Calculation of thermal processes in laser welding

Grigor' yants, A. G. et al. *Izv. V. U. Z. Mashinostroenie*, (11): 135-138, 1981. (In Russian)

To describe the heat propagation in the sheet a non-linear heat conductivity equation is solved by a finite-difference method, taking into account the temperature dependences of the thermo-physical characteristics of the metal. Taking as an example the laser welding of Ti-alloy VT28 sheet it is shown that the use of such a non-linear mathematical model provides more exact results than the classical scheme, particularly at temperatures greater than 800K.

3.5 (5) Tuning of a parametric model for the laser cutting of steels

Esposito, C. and Daurelio, G. (Centro Laser, Italy). *Optic Lasers Engineering*, 2(3): 161-171, 1981.

Examines the mathematical models developed to describe the gas jet laser cutting process. Some experimental results for the laser cutting of carbon and stainless steels are also shown. These results are then used to 'tune' the theoretical models in order to obtain a method for making predictions about the parameters (4 refs.)

3.5 (6) Stability and symmetry in inertial confinement fusion

Emery, M. H., et al. Rept. No. AD-A122 421/1 (NRL-MR-4947), 18pp., December 1982. Naval Research Laboratory, Washington, DC, USA).

The asymmetries of spherical implosions driven by direct laser illuminations are of fundamental concern to the inertial confinement fusion community because they provide severe limitations on high gain pellet designs. Theoretical progress on several fronts has recently been made through numerical simulations in providing a more complete understanding of the physical processes involved in these asymmetries and instabilities. The results also suggest methods of controlling these processes and their implications for laser fusion systems design. Stability and symmetry issues have been investigated. Laser matter coupling and scaling laws relating the asymmetry results to spherical pellet designs have been investigated.

3.5 (7) Mathematical modeling of laser material interactions

Steen, W. M. and Mazumder, J. (Imperial College, UK). Final Rept. No. AD-A141 096/8, 92pp., November 1983.

Describes the development of a finite difference model of laser/material heat transfer, which allows for thermal conduction, surface convection and radiation, variation in thermal properties, latent heat, keyhole formation and convection in the melt pool.

3.5 (8) Optimising a laser-welding regime for a heat resistant nickel alloy

Fedorov, B. M. et al. *Avtom. Svarka.*, (1): 48-50, 1983. (In Russian)

The significance of calculated parameters was studied in a multifactorial mathematical model for laser welding of 1.5mm thick KhN68VMTYuK Ni alloy samples. The parameters included welding speed, arc power and degree of focusing. It was found that increasing welding speed at constant radiation capacity decreased the threshold radiation energy and the molten area. Increasing radiation capacity at constant welding speed increased energy density, while the latter decreased with deviation of the focal plane from the sample surface. Therefore, welding speed and radiation capacity must be maximum and the focus must be buried in the part being welded to provide melt-through equivalent in width to the minimum cross-sectional area.

3.5 (9) Absorption measurements for high power laser material processing

Gay, P. and Manassero, G. (Fiat Research Centre, Italy). In, *ICALEO '83 Proc. Materials Processing Symp.*, Vol. 38, 14-17 November 1983, Los Angeles, CA, (Ed. E. A. Metzbower), pp. 224-228. Laser Institute of America, Toledo, OH, USA, 1984.

Laser absorption measurements at surface temperatures lower and higher than melting point were carried out. The data, obtained by fitting temperature measurements with a bidimensional mathematical model, have been obtained for power densities of $10^7 W/m^2$ at 5kW laser power. A significant drop of absorption for melted surface was observed. (8 refs.)

3.5 (10) A direct modeling of the thermal conduction on laser machining

Tosch, R., Gruber, H. and Fritzsche, K. (Tech. Hochschule Leipzig, Germany). *Feingeraetetechnik*, 33(3): 119-120, 1984. (In German)

A numerical model is presented which describes thermal conduction in laser irradiated materials considering the temperature dependent thermophysical and optical quantities.

3.5 (11) Numerical modeling of laser material processing

Covle, R. J. and Rajaram, S. (Western Electric, Priceton, NY). In, *ICALEO '83, Proc. Materials Processing Symp.*, Vol. 38, 14-17 November 1983, Los Angeles, CA, (Ed. E. A. Metzbower), pp. 216-223. Laser Institute of America, Toledo, OH, USA, 1984.

Many of these applications require a detailed understanding of laser material interactions to control the weld penetration and extent of the heat affected zone (HAZ). To gain a better understanding of the welding process, it is desirable to generate an accurate mathematical model which can predict material behaviour and temperature profiles for various sets of laser parameters. This study was initiated to develop a model which could be applied to relatively low-power continuous wave welding (several hundred watts) of thin sections. Metallographic inspection of weld cross-sections was used to determine the effectiveness of the model in predictions of weld zone geometry of Nitronic 60 stainless steel. (8 refs.)

3.5 (12) Thermal modeling of laser welding with a quasi-one-dimensional approach

Blottner, F. G. (Sandia National Laboratories, USA). In, *Modeling of Casting and Welding Process 11*, 31 July – 5 August 1983, Henniker, NH, USA, pp. 349-361. Metallurgical Society/AIME, Warrendale, PA, USA, 1984.

A relatively simple numerical technique for predicting the melt depth is developed for pulsed laser welding. The two-dimensional aspects of the solution are approximated with an effective area in the one-dimensional energy equation. The enthalpy method is used with an implicit numerical scheme which provides a rapid solution procedure. A preliminary model of metal absorptivity of the laser radiation is developed but requires some experimental input.

3.5 (13) Laser welding – a parametric study on inco718

Mehta, P., Cooper, E. B. and Miller, R. (GEC, USA). In, *ICALEO '84, Proc. Materials Processing Symp.*, Vol. 44, 12-15 November 1984, Boston, MA, (Ed. J. Mazumder), pp. 43-52. Laser Institute of America, Toledo, OH, USA, 1985.

Statistically designed experimentation can be a powerful tool to facilitate the collection of data in process capability/exploration studies. Statistically designed experimentation consists of an iterative approach for the collection and evaluation of data. The payoff from such an approach is the development of a (statistical) process model with a minimal amount of experimentation. This paper describes the philosophy, approach, and results of a statistically designed experiment conducted to explore a laser welding process. (7 refs.)

3.5 (14) Solution of a Stefan problem in the theory of laser welding by the method of lines

Davis, M. and Kapadia, P. (Essex University, UK). *J. Comput. Phys.*, 60(3): 534-548, 1985.

Use of a laser beam as the source of energy for penetration welding gives rise to a long, thin cylindrical hole surrounded by molten metal. Material moves from the front to the rear of the hole as the workpiece is translated relative to the laser, by flowing around the hole. A computer program has been written which solves the equations governing a 2-dimensional steady-state mathematical model. Computed results have been found to agree low speeds of welding, and at higher speeds they give results similar to what is observed in practice.

3.5 (15) Temperature distribution in a two-layer disk composed of a metal and a coated film with laser irradiation

Ohmura, E. (Osaka University, Japan). *Heat Transfer – Jpn. Res.*, 14(3): 32-46, 1985.

Theoretical equations were derived by both Laplace and finite integral transforms. In this model, the intensity distribution of the laser beam was symmetrical with respect to the central axis and the beam was absorbed in the interior of the coated film. (7 refs.)

3.5 (16) Theoretical model of oxygen assisted laser cutting

Schuocker, D. et al. (Technical University Wien, Austria). In, *Proc. 5th Int. Symp. on Gas Flow and Chemical Lasers*, 20-24 August 1984, Oxford, UK, (Eds. A. S. Kaye and A. C. Walker), pp. 111-116. Adam Hilger, Bristol, UK, 1985.

At the 4th GCL in Stresa, Italy an improved theoretical model of laser cutting was presented. That model is based on the assumption that the momentary end of the cut a nearly vertical surface, is covered by a thin molten layer, that loses mass by evaporation and by ejection of liquid material due to the friction between the reactive gas flow and the molten material. Two final equations, the energy balance and the mass balance of the molten layer, are obtained and allow the determination of all relevant quantities. These equations have been extended by time dependent terms.

3.5 (17) Three-dimensional model for convection in laser melted pool

Chan, C., Mazumder, J. and Chen, M. M. (University of Illinois, USA). In, *ICALEO '84, Proc. Materials Processing Symp.*, Vol. 44, 12-15 November 1984, Boston MA, (Ed. J. Mazumder), pp. 17-27. Laser Institute of America, Toledo, OH, USA, 1985.

A three-dimensional axis-symmetry model of the fluid flow and heat transfer of laser melted pool is developed. The model corresponds physically to a stationary laser source. Non-dimensional form of the governing equations is derived. Four dimensionless parameters arise from the non-dimensionalisation, namely, Marangoni number (Ma), Prandtl number (Pr), Dimensionless melting temperature (T_m^*), and Radiation factor (RF). Their effects and significances are discussed. Numerical solutions are obtained. The solid liquid interface, which is not known a priori, is solved. Quantitative effects of the dimensionless parameters on pool shape are obtained. In the presence of the flow field, the heat transfer becomes convection dominated. Its effect on isotherms within the molten pool is discussed. Experimental results are also obtained and presented.

3.5 (18) A two-dimensional model for mass transport in laser surface alloying

Chande, T. and Mazumder, J. (University of Illinois, USA). In, *ICALEO '84, Proc. Materials Processing Symp.*, Vol. 44, 12-15 November 1984, Boston, MA, (Ed. J. Mazumder), pp. 140-151. Laser Institute of America, Toledo, OH, USA, 1985.

The process of alloy generation during laser surface alloying was examined. A numerical model was developed that solved the two-dimensional, transient equation for convection-diffusion of matter in the melt pool using the alternate-diagonal implicit method. Velocity distributions used were for steels for a laser power of 1.5kW, a 0.0001m beam diameter and a traverse speed of 0.01m/s. Calculations were made for a uniform surface mass flux of 5 and $10kg/m^2$-s and values of 10 and 100 for the Peclet number for diffusion. The results showed that powder particles injected into the melt pool melted practically instantaneously, that good mixing was determined by the pattern of fluid flow, that average solute contents increased linearly with increasing interaction time and that a change in the solute diffusivity by a factor of 10 did not greatly alter the predicted solute distribution. Good mixing was also found with surface flux over only a part of the melt pool. The conclusions were that fluid flow dominates the process of mass transport and determines the nature of the solute distribution and that powder *feed is an effective method for surface alloying. (11 refs.)*

3.5 (19) Two-dimensional modeling of conduction-mode laser welding

Russo, A. J., et al. (Sandia National Laboratories, USA). In, *ICALEO '84, Proc. Materials Processing Symp.*, Vol. 44, 12-15 November 1984, Boston, MA, (Ed. J. Mazumder), pp. 8-16. Laser Institute of America, Toledo OH, USA, 1985.

The development of a predictive code for conduction-mode laser welding requires that models for the temperature-dependent absorption coefficient and vapourisation interactions should be included with the thermal conduction calculation. An attempt to understand and quantify these surface phenomena by comparison of a two-dimensional code with experimental results is described.

3.5 (20) Analytical investigation of thick film ignition module soldering by laser

Chang, D. U. (Control Laser Corp, USA). In, *Laser Welding, Machining and Materials Processing – Proc. Int. Conf. on Applications of Lasers and Electro-Optics, ICALEO '85*, 11-14 November 1985, San Francisco, CA, (Ed. C. Albright), pp. 27-38. Laser Institute of America, Toledo, OH, USA/IFS (Publications) Ltd, Bedford, UK, 1986.

A study of the effects of process parameter variations in laser soldering of thick film ignition module is presented in this paper. The process parameter study was done with the use of an analytical model. The model was compared with the experimental data reported earlier. The effect of beam power, beam on-time, beam spot diameter, preheat temperature, and the materials' thermal properties on soldering is discussed. The model was found useful for better understanding and control of the laser soldering process. (14 refs.)

4

LASER PROCESSES

4.1 LASER WELDING

4.1.1 General

4.1.1 (1) High energy density beam welding

In, *Proc. Int. Symp. of Japanese Welding Society – Advanced Welding Technolgy*, Session 1, 25-27 August 1975, pp. 3-162. Japanese Welding Society, Tokyo, Japan, 1975.

Session 1 consists of 26 papers dealing with electron beam welding, covering problems such as laser and electron beam welding of engineering components; effect of electron and laser beam distribution on weld seam formation; welding of high-strength aluminium alloy; and welding of titanium to carbon steel using an insert metal. Also includes the subject of laser and light beam welding, such as evaluation of a high-power welding CO_2 laser, wall-focusing effect of laser beams, economic cutting and welding with the CO_2 laser, fusion zone purification in high-power CO_2 laser welding, and development of the light beam welding process.

4.1.1 (2) Laser spot welding and real-time evaluation

Saifi, M. A. and Vahaviolos, S. J. (Western Electric Co. USA). *IEEE J. Quantum Electron.*, (5th Biannual Conf. on Laser Engineering and Applications, 28-30 May 1975, Washington, DC, USA), QE-12(2) part 2: 129-136: 1976.

4.1.1 (3) Laser welding – the present state-of-the-art

Breinan, E. M., Banas, C. M. and Greenfield, M. A. (Wright-Patterson Air Force Base, USA). Tech. Rept. No. AFML-TR-149, 73pp., November 1975.

An early review of the evolution of laser

systems from the first successful operation of the ruby laser in 1960 to 1975. Includes discussion of the then available high-power laser equipment and applications, laser welding performance, and weld characteristics, and the properties of laser welds in metals and alloys.

4.1.1 (4) Limited penetration laser welding applications

Bolin, S. (Raytheon Co, USA). SME Tech. Paper Ser. AD 1975, 19pp. (Presented at Assemblex Conf., 7-9 October 1975, Rosemont IL, paper AD75-753. Society of Manufacturing Engineers, Dearborn, MI, USA.

Applications are described as well as material properties and their interacting relationships in successful fusion joining. The inherent ease of automating the laser industrial process has made it a cost-effective choice.

4.1.1 (5) Angular deformations during laser welding

Gol'tsova, V. P. et al. *Autom. Weld.*, 30(2): 35-37, 1977.

4.1.1 (6) Current status and future prospects of special welding techniques

Chene, J. J. (Technical Hochsch, Switzerland). *Schweisstech Soudure*, 67(4): 83-92, 1977. (In German and French)

Discusses special welding techniques and their applications and compares them: electron beam welding, laser welding, high-temperature brazing, diffusion welding, friction welding, and explosion welding.

4.1.1 (7) Design for laser beam welding

Shewell, J. R. (Lasertool Consult, USA). *Weld Des Fabr.*, 50(6): 106-110, 1977.

Laser welding processes, conduction-limited and deep penetration, use only the parent metal with no added filler. In conduction-limited laser welding, the more common, the metal absorbs the laser beam at the work surface. The subsurface region is heated entirely by thermal conduction. The term laser welding normally refers to this method, which uses solid state and moderate power CO_2 lasers. The second type, deep penetration welding, requires a high power CO_2 laser. Thermal conduction does not limit penetration. Laser beam energy is delivered throughout the depth of the weld, not just to the top surface.

4.1.1 (8) Present state of continuous laser welding technology

Velichko, O. A. et al. (Weld Inst E. O. Paton, USSR). *Avtom. Svarka.*, (5): 44-50, 1977. (In Russian)

Early state-of-the-art report. (39 refs.)

4.1.1 (9) Laser welding

Arata, Y. and Miyamoto, I. *Technocrat*, 11(5): 33-42, 1978.

Emphasises laser welding advantages such as high energy density; gain control at a convergence point; use under atmospheric conditions; time-sharing processing at a high speed with high accuracy; and the most suitable heat source for automatic control. Disadvantages include a dependency on the surface condition of the workpiece, and expensive equipment. Weld depth lies between those obtained by electron beam welding in air and in vacuum. Applications of laser welding include electronic parts for nuclear plant equipment, seam welding of vanes for gas turbine engines, pulsed lap welding of Zr cases for nuclear reactor fuel rods, and welding of automotive transmission gear boxes with minimum thermal distortion.

4.1.1 (10) Positioning system with laser for automatic micro spot laser welding

Ono, A. et al. (Toshiba Corp, Japan). In, *Ann. CIRP*, 28(1) 1979, Manuf. Technol, 29th General Assembly of the CIRP, Davos, Switz, 26-August 1 September 1979. Publ. by Tech Rundsch, Berne, Switz, 1979, pp. 317-320.

The positioning system using a He-Ne laser was developed for spot YAG laser welding. Non-contact, accurate, and rapid positioning was attained by detecting the reflection beam from the edge of the part close to the welding spot with the He-Ne laser beam scanned over the edge of the part. After detecting the edge, the

YAG laser welding optical system moves a given distance from the edge to the accurate welding spot, and then the YAG laser irradiates the welding spot. The YAG laser welder is used for mass production of high-quality electric devices.

4.1.1 (11) Solid state lasers in welding

Daene, K. (Zentralinst fuer Schweisstech der DDR, Germany). *Schweisstechnik*, 29(1): 13-16, 1979. (In German)

Describes three different solid state laser units which can be used for welding applications. Instructions for weld preparation in laser welding are presented, and technological conditions characteristic of laser welding and drilling are discussed.

4.1.1 (12) Source book on electron beam and laser welding

Schwartz, M. M. (American Society for Metals. Metals Park, OH, USA, 1980, 398pp.

Compilation of 37 articles previously published in journals and proceedings covering all important aspects of electron beam and laser welding. For laser welding, the state-of-the-art, design and tooling, applications in power plants and automotive industries are reviewed. Mechanical properties of laser welded high-strength steels are treated in the later sections.

4.1.1 (13) Special features of formation of the weld in continuous laser welding

Avramchenko, P. F., Shovkoplyas, V. M. and Velichko, O. A. *Automat, Weld.*, 33(4): 43-45, 1980.

4.1.1 (14) Improving weld quality with laser monitoring

Burris, M. K. (Bendix Corp, USA). Rept. No. BDX-613-2497, 20pp. October 1980. Bendix Corp, Kansas City, MO, USA.

Describes the development of a CO_2 laser beam monitoring system to enhance the quality of a weld by monitoring equipment parameters affecting the weld.

4.1.1 (15) Increasing energy absorption in laser welding

Jorgensen, M. *Met. Constr.*, 12(2): 88, 1980.

The effect of oxygen concentration on the energy absorption and penetration depth in laser welding of mild steel was determined. Welding depths were measured at gas flow velocities of 11, 14.5, and 17mm/s. It was found that gas velocity had no significant effect but an increase in depth and a decrease in reflectivity was obtained with small 0 additions.

4.1.1 (16) What's new in electron beam and laser welding

Brock, T. *Mod. Mach. Shop*, 52(12): 80-89, 1980.

The laser beam welder, unlike the electron beam welder, has no size limit and can be bent with mirrors for welding. It is unaffected by a magnetic field and is used to weld contacts to electromagnets. The laser beam is used to weld plastic and seal containers affected by a vacuum.

4.1.1 (17) Arc augmented laser welding – process variables, structure and properties

Alexander, J. and Steen, W. M. In, *Proc. the Joining of Metals: Practice and Performance*, Vol. 2, 10-12 April 1981, Coventry, UK, pp. 155-160. Institution of Metallurgists, London, UK, 1981.

An electric arc can be stabilised by a laser generated plasma and the energy of the arc used to augment that of the laser. For bead on plate welds, it is possible to use up to 100A of arc current without adversely affecting the weld bead profile. This can result in an increased welding speed on that of the laser alone or an increased penetration. Successful butt welds can be produced using this technique, although in some welds there was an increase in porosity not generally observed in either the arc or laser alone welds. (5 refs.)

4.1.1 (18) Laser welding

Metzbower, E. A. (US Naval Research Laboratory, USA). *Naval Engineering J.*, 93(4): 49-58, 1981.

Discusses lasers, laser welding; compares laser welding to other welding techniques; gives a synopsis on how to use the laser to weld; compares the properties and structures of laser welds to welds fabricated by other processes and, finally, indicates the potential for lasers in other production engineering applications.

4.1.1 (19) Laser welding of metal: effect of travelling velocity on weld shape

Arata, Y. et al. *J. Japanese Welding Soc.*, 50(4): 404-409, 1981. (In Japanese).

Correlation between the travelling velocity and the dimensions of the stainless steel laser weld bead is examined over the range 0.2-200m/min. Interaction between the cavity and the incident beam is analysed. The penetration depth and the joining rate are estimated on the basis of this correlation and on heat conduction principles. With fixed focusing, cyclic change occurs in the beam intensity at the bottom of the cavity, producing a step-like change in the dimensions of the bead at different velocities. Under optimal focusing conditions the ratio of penetration depth: mean bead width is practically constant over a wide range of travelling speeds. (8 refs.)

4.1.1 (20) Special features of welding with an inclined beam

Bashenko, V. V. et al. *Weld Prod.*, 28(7): 12-14, 1981.

In beam welding, the weld pool is screened by the breakdown products. Consequently, the maximum thermal efficiency is achieved if the beam is deviated by an angle comparable with the angle of opening of the vapour-gas channel. In this case, the depth of penetration in the pulsed conditions increases by 22% in laser welding.

4.1.1 (21) Fabrication of laser seal-welded targets for particle beam fusion experiments

Armstrong, S., Flick, F. F. and Perry, F. (Los Alamos National Laboratory, USA). *J. Vacuum Science Technology*, 24(4): 1085-1086, 1982. (Proc. National Symp. of American Vacuum Society, Pt. 2, 2-6 November 1981, Anaheim, CA)

Proposed particle beam fusion targets require seamless spheres filled with high pressure deuterium-tritium gas. Conventional fill techniques using small diameter fill tubes produce perturbations upon implosion. These perturbations may cause instabilities and jetting of target material into the fuel core, resulting in a low target yield. A laser fabrication technique has been developed to eliminate the need for fill tubes, by seal-welding a small fill hole in the target wall, while in a high pressure gas environment. The fabrication parameters for processing simulated laser seal-welded targets are discussed in detail. These steps include electroplating, micromachining and laser welding.

4.1.1 (22) Optimising the position of the focal plane in laser welding

Volobuev, Yu. V. et al. (N. E. Bauman Technical College, USSR). *Autom. Weld*, 35(4): 47-49, 1982.

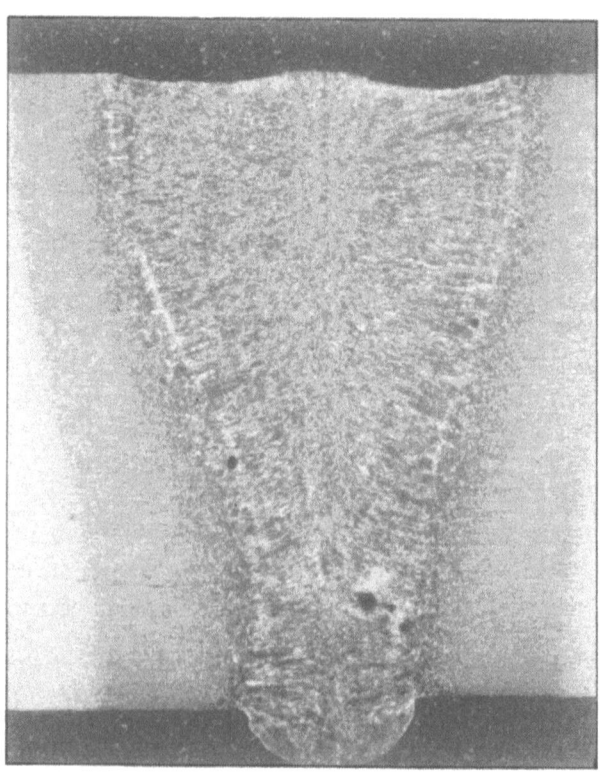

Laser weld made using wire feed to fill a 2mm gap in 8mm thick C-Mn steel plate (See 4.1.1 (26))

4.1.1 (23) Proc. 30th national conference – Australian welding institute, welding technology '82.

9-15 October 1982, Tasmania, Australia. Australian Trade Publications Pty. Ltd. Chippendale, NSW, Australia, 1982, 344pp.

Forty-one papers discuss general aspects of welding including the development of new welding processes, such as Innershield, the comparison of fluxed-cored-arc-welding (FCAW) and manual-metal-arc-welding (MMAW) processes and the selection of a shielding gas and its effect on weld metal analysis. Robotic spot welding techniques in the Japanese automobile industry and the characteristic features of high-power laser welding are evaluated.

4.1.1 (24) Acoustic inspection of the penetration depth of metal in pulsed laser welding

Korlyakov, V. K. *Weld. Prod.*, 30(9): 32-33, 1983.

An experimental examination of acoustic emission (AE) revealed the integral values of the high- and low-frequency components of emission. A method of determining the penetration depth of metal using the information-carrying parameters of AE which is not affected by technological interference is proposed.

4.1.1 (25) Laser welding

Mazumder, J. (University of Illinois, USA). In, *Mater. Process Theory Practice, Vol. 3, Laser Mater. Process.*, pp. 113-200, North-Holland Publishing Co., Amsterdam and New York, 1983.

The basic principles of laser welding and laser welding variables are outlined. Describes various commercial CO_2 laser systems having powers from a few watts to 15kW available for a multitude of laser processing applications. High-power (up to 400W average CW power) pulsed Nd:YAG lasers are also available for overlapping spot welding applications. Due to its shorter wavelength (1.06μm) the Nd:YAG laser can produce higher power density compared to a CO_2 laser. Otherwise, welding principles for Nd:YAG lasers are the same as CO_2 lasers.

4.1.1 (26) Laser welding – techniques and testing

Watson, M. N., Oakley, P. J. and Dawes, C. J. (The Welding Institute, UK). In, *Proc. 1st Int. Conf. on Lasers in Manufacturing*, 1-3 November 1983, Brighton, UK. (Ed. M. F. Kimmitt), pp. 133-142. IFS (Publications) Ltd, Bedford, UK, 1983. Also in *Metal Construction*, 17(5): 288-291, May 1985.

The nature of the high power CO_2 laser welding process requires the development of techniques to measure the process parameters and also puts new requirements on the techniques used for assessment of weld quality and properties. This paper describes, with practical examples as illustration, the measurement of laser power, focus position, and other welding variables. Also described is the evaluation of laser welded joints in various materials of engineering interest up to 13mm in thickness (welded with laser powers up to 6kW), by visual, radiographic and metallographic examination, and the testing of tensile, fracture toughness and fatigue properties, again with practical examples. (20 refs.)

4.1.1 (27) A survey of the present state of automation and advanced technology fusion welding in Japan and the European community

Jones, S. B. and Russell, J. D. Commission of the European Communities, Luxembourg, 1986

Presents a management summary and conclusions from a study which considered aspects of the development of fusion welding in Japan and the European Community with the objective of indicating areas where further European action would be justified. The level of technical achievement in European and Japanese research into arc welding processes, industrial robotics, EB and laser technology is assessed and found to be generally equivalent. The Japanese, however, have paid particular attention to the practical realisation of research ideas as industrial products. This approach has given them commercial advantages in several technical areas, including low-cost, advanced welding power sources. *(82 refs.)*

4.1.1 (28) Continuous wave and pulsed periodic regimes for welding with an electroionisation CO_2 laser

Basov, N. G. et al. (Academy of Science, USSR). *Bulletin Acad. Sci. USSR, Phys. Ser.*, 48(12): 19-29, 1984.

Investigates welding with pulsed periodic laser radiation over a wide range of pulse lengths and pulse repetition frequencies for an average radiation power up to 3kW and compares it with welding by continuous wave laser radiation. The possibility of varying the parameters of laser radiation over such wide limits is offered by the use of a universal electroionisation CO_2 laser with grid control of the electron gun current. An analysis of the experimental results obtained provided the possibility of determining the dependence of the melting efficiency, the melting capacity, and the seam-shape factor on the welding rate and the parameters of the laser radiation in the continuous wave and pulsed periodic regimes.

4.1.1 (29) Effect of surface condition and edge preparation on penetration in laser welding

Bashenko, V. V. et al. *Weld. Prod.*, 31(5): 27-29, 1984.

It has been found that optical breakdown taking place in argon and CO_2 in the weld metal absorbs 45-60% of laser radiation energy. By varying the composition of the vapour phase in laser welding by introducing various additions, it is possible to alter the effect of radiation on the metal; in particular, efficiency of the process can be increased by depositing potassium nitrate flux into the joint coatings.

4.1.1 (30) Special features of laser welding with complete penetration

Skripchenko, A. I., Surkov, A. V. and Bashenko, V. V. *Weld. Prod.*, 31(5): 24-26, 1984.

Suggests that the diameter of the upper part of the puddle formed in laser welding with complete penetration is considerably greater than the diameter of the laser beam. The upper part of the channel is heated by the laser plasma, whereas the lower part is heated by the radiation

which is placed through the plasma. Authors claim that by varying the power of transmitted radiation, it is possible to stabilise the process of complete penetration and also predict and control the shape of the penetration zone.

4.1.1 (31) Vacuum laser welding method

Arata, Y. and Oda, T. *J. High Temp. Soc. Jpn.*, 10(1): 24-27, 1984. (In Japanese)

To supress laser plasma in high power CO_2 laser welding, a new method of vacuum laser welding was developed applying evacuation of the atmosphere gas environment by developing a special aerodynamic window.

4.1.1 (32) Welding processes and applications – future trends (resistance, electron beam and laser welding)

Muthukrishnan, S. (United Nations Industrial Development Organisation, Austria). Rept. No: UNIDO-ID/WG.420/14, 72pp., April 1984.

Laser welding is discussed including its technical aspects and applications control methods, processing variables, component and joint design.

4.1.1 (33) Basics of laser welding

Darchuk, J. M. and Migliore, L. R. (Spectra-Physics, USA). *Lasers Appl.*, 4(3): 59-66, 1985.

4.1.1 (34) Fundamental phenomena during vacuum laser welding

Arata, Y. et al. (Osaka University, Japan). In, *ICALEO '84, Proc. Materials Processing Symp.*, Vol. 44, 12-15 November 1984, Boston, MA, (Ed. J. Mazumder), pp. 1-7. Laser Institute of America, Toledo, OH, USA, 1985.

Observation was performed with a high-speed camera and by the transmission X-ray method. It was found that vacuum laser welding can almost completely suppress laser plasma and that this allows deep penetration at a very slow welding speed. Under these conditions, the shape and behaviour of the beam hole during welding were very similar to electron beam welding. The fundamental characteristics of vacuum laser welding were also studied, including the effect of gas pressure and the welding speed on the penetration depth. The penetration depth increased with decreasing pressure and also as the welding speed decreased. A penetration depth of over 40mm was achieved at a power of 11kW, a pressure of 10^{-3}Torr, and a speed of 10cm/min. Vacuum laser welding using an aerodynamic window was proposed for practical applications, and a penetration of over 25mm was subsequently achieved, even at a pressure of 50Torr.

4.1.1 (35) In-process acoustic emission monitoring of laser welds

Whittaker, J. W. et al. (Oak Ridge Y-12 Plant, USA). Rept. No: Y/DW-592 CONF-851162-2, 7pp; July 1985.

Acoustic emission monitoring systems were developed for monitoring in-process pulsed and continuous wave laser welds. Thin sheet, coupon welds and very small, complex geometry component welds were monitored. The results indicate that a linear relationship exists between AE energy production and weld penetration in pulsed welds. Preliminary results indicate that beam/seam misalignment can be determined from the AE generated during welding.

4.1.1 (36) Laser and electron beams for deep, fast welding

Kuvin, B. F. *Weld. Des. & Fabr.* 58(8): 34-40, 1985.

Two aims drive implementation of laser and electron beam welding; higher welding speed and reduced distortion of thin parts. Lasers and electron beams satisfy both aims and produce welds with high depth-to-width ratios, to 40:1, and narrow heat-affected zones. Distortion is minimised by their highly localised heat input.

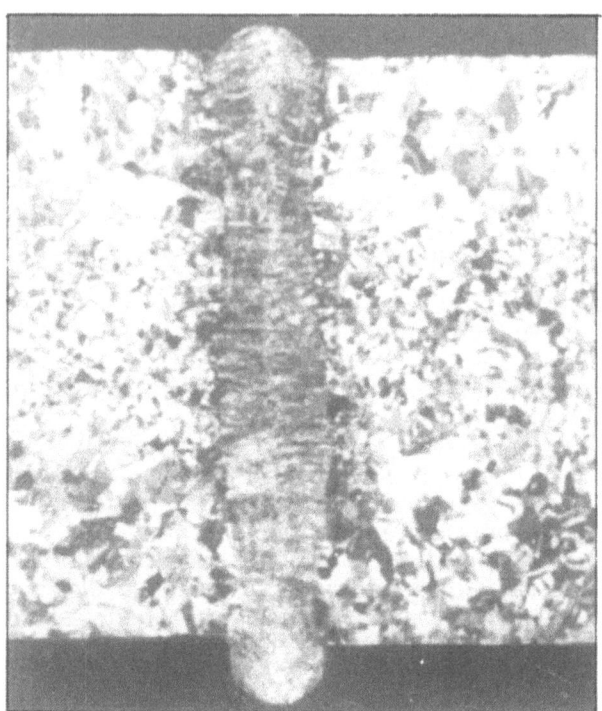

Macrography of LB and EB welds (See 4.1.1 (38))

4.1.1 (37) Some factors effecting penetration in laser welding

Moon, D. W. (Naval Research Laboratory, USA). In, *ICALEO '84, Proc. Materials Processing Symp.*, Vol. 44, 12-15 November 1984, Boston, MA, (Ed. J. Mazumder), pp. 53-59. Laser Institute of America, Toledo, OH, USA, 1985.

In an effort to optimise penetration in high power laser beam welding the following four aspects have been investigated: (1) the relationship between the external· telescope positions and image points; (2) weld penetration as a function of the nozzle distance of the plasma suppressing gas; (3) penetration vs nozzle size relationship; and (4) the effect of the shield's height on penetration at various power levels.

4.1.1 (38) What choice for high integrity joints: electron beam or laser beam welding?

Sayegh, G. (Sciaky SA, France). In, *Proc. 2nd Int. Conf. on Lasers in Manufacturing*, 26-28 March 1985, Birmingham, UK, (Ed. M. F. Kimmitt), pp. 11-22. IFS (Publications) Ltd, Bedford, UK, 1985.

Through a few industrial application examples, specific characteristics and advantages of high energy density beams (EB and LB) are illustrated, thus justifying the use of these processes in production. Technical performance comparison of the two processes and economical considerations, results in the definition of three types of applications: thin gauge material for which LB is more appropriate; thick gauge material for which EB is more appropriate; and medium gauge material for which effective and strong competition exist between LB and EB.

4.1.1 (39) Development of an automatic laser welding machine to produce gas filled capsules

Stevenson, P. (Frazer-Nash Ltd, UK). In, *Proc. 3rd Int. Conf. on Lasers in Manufacturing*, 3-5 June 1985, Paris, France, (Ed. A. Quenzer) pp. 157-167. IFS (Publications) Ltd, Bedford, UK, 1986.

An automatic machine to produce temperature sensing elements was required to replace four ageing electric seam welders. The characteristics of CO_2 laser welding had been shown to be superior in improving weld speed, quality and accuracy of weld position. During development, problems of gas leaking through weld imperfections arose and were subsequently found to be due to traces of capsule gas absorbing laser power and internal capsule pressure disturbing the weld pool. These problems were subsequently overcome by modifying the laser delivery system and the method of clamping the components.

4.1.2 Laser welding of metals

4.1.2.1 General

4.1.2.1 (1) Fusion welding of thin metal foils

Casey, H. (Los Alamos Science Laboratory, USA). SME Tech. Paper Ser. AD75-869, 12pp., Advanced Welding and Brazing Workshop, 28-30 October 1975, Chicago, IL, USA). Society of Manufacturing Engineers, Dearborn, MI, USA.

Reviews some of the problems associated with such welding and describes current techniques employed to join foils. Techniques for fusion welding 0.25mm thick foils of copper, aluminium and stainless steels have been developed using laser welding equipment. These techniques, together with the related aspects of joint design, tooling and fixturing, joint preparation and some necessary modifications to the commercially available welding equipment, are included.

4.1.2.1 (2) Guide to laser metal welding – to 1500 watts

Engel, S. L. (GTE Sylvania, Inc, USA). SME Tech. Paper MR75-579, 17pp., December 1975. Society of Manufacturing Engineers, Dearborn, MI, USA.

Demonstrates that the types of metals to be welded and the penetration required determine the minimum size of laser required. Identifies an optimum focal point position to be below the surface of the material. Fit-up and alignment tolerences are discussed and specified as functions of weld penetration, bead width and joint geometry. Selection of materials is discussed and illustrated through several examples. Safety measures are emphasised. (4 refs.)

4.1.2.1 (3) Bright spot for pulsed lasers

Bolin, S. R. (Raytheon Co, USA). *Weld. Des. Fabr.*, 49(8): 74-77, 1976.

Low energy input, confined high intensity, and non-contact are characteristics which have made pulsed laser welding useful where precision parts are to be joined, contamination might be a problem, heat distortion cannot be tolerated, or welding must be done within a glass enclosure from the outside. The pulsed laser delivers pulses of energy at rates up to 200 per second. The machine makes high-quality spot welds as deep as 0.080 inches. Shallower seam welds can be made at rates over 2 inches per second.

4.1.2.1 (4) Influence of gap size on the formation and strength of a spot lap joint in laser welding

Baronov, M. S. (All-Union Correspondence Institute of Mechanical Engineering, USSR). *Weld. Prod.*, 23(5): 19-21, 1976.

In the laser welding of spot lap joints, a gap between the plates and partial penetration of the bottom plate are permissible. This gap size is governed by the area of fusion and the joint strength decreases as the gap increases, as a result of reduction of the area of fusion and of the depth of penetration of the bottom plate. Deflection of the surface of the weld pool metal does not have any large influence on shear strength.

4.1.2.1 (5) Kilowatt welding with a laser

Engel, S. L. (Caterpillar Tractor Co, USA). *Laser Focus*, 12(2): 44, 46-47, 50-53, 1976.

Outlines the benefits of laser welding including its simplicity and the fact that the same optical system, gas nozzle and shield gas can handle most metals from carbon steel to zirconium alloys. Furthermore, the laser welding process changes little with the material being worked.

4.1.2.1 (6) Role of shielding gas in high power CO_2 (CW) laser welding

Seaman, F. D. (ITT Research Institute, USA). SME Tech. Paper Ser. MR77-982, 12pp, 1977. Society of Manufacturing Engineers, Dearborn, MI, USA. Presented at Latest Uses of Lasers in Manuf. Conf., 27-29 September 1977, Los Angeles, CA.

Laser welds must be shielded from air, but shielding gases have been observed to react with the laser beam, robbing power from the weld. Cross sections of welds made with various shielding gases and gas mixtures show a 60% difference in penetration. However, gases that permit the greatest penetration do not blanket the weld effectively at the characteristically high laser welding speeds.

4.1.2.1 (7) Future development of continuous laser welding

Crafer, R. C. In, *Proc. Conf. on Welding in the Aerospace Industry – Design, Materials, Welding Methods, Maintenance, 7-8 December 1978, Berlin, W. Germany, pp. 64-68.* Deutscher Verlag fur Schweisstechnik (DVS) GmbH, Dusseldorf, W. Germany, 1978.

Describes the welding process in detail, from experience gained on both 2 and 5kW machines, and attempts to answer those questions which a professional welding engineer might consider relevant to process evaluation. Materials welded include: austenitic stainless steels, Al-base alloys and Ti-base alloys.

4.1.2.1 (8) How to make better laser welds

Engel, S. L. (HDE Systems, USA). *Weld. Des Fabr.*, 51(1): 62-65, 1978.

The same laser optical arrangement, gas nozzle, and shield gas will handle most metals and alloys from aluminium to zirconium. Laser welding works best on thin metal parts. Metal and depth of weld determine the minimum power required by the laser. Most joint geometries are suitable for laser welding, but the laser beam focuses to a tiny spot, calling for close tolerances and alignment.

4.1.2.1 (9) Improved welding performance from a 2kW axial flow CO_2 laser welding machines

Crafer, R. C. In, *Proc. Advances in Welding Processes Conf.*, Vol. 1, 9-11 May 1978, Harrogate, UK, pp. 267-278. Welding Institute, Cambridge, UK, 1978.

Mild and stainless steels have been welded faster and deeper than previously was thought possible, and high conductivity materials such as Al alloy have been welded at lower power levels than hitherto considered feasible. All these materials exhibit optimum joining rate in the thickness range with extended thick section performance in the steels as a result of efficient control of the metal vapour plasma by means of the plasma jet attachment. (7 refs.)

4.1.2.1 (10) Laser spot welding of metals: operation and problems involved

Kullen, J. (University of Stuttgart, W. Germany). *WT Z Ind Fertigung*, 68(1): 13-17, 1978. (In German)

Describes significant characteristics of laser beam welding; suitable pulsed laser welding units are reviewed, and examples are presented of laser welding applications in watchmaking, precision engineering, and electrical engineering practice.

4.1.2.1 (11) New assembly techniques using laser welding

Seaman, F. D. (Society of Manufacturing Engineers, USA). SME Tech. Paper AD79-895, 18pp., 1979. Society of Manufacturing Engineers, Dearborn, MI, USA.

The welding of thick steel sections and the high-speed assembly of hard to weld sheet metal are described. High beam power is used for heavy place of 1:45% Mn steel and low beam power for sheet metal of Al, carbon and stainless steels.

4.1.2.1 (12) Process aspects of laser welding

Crafer, R. C. (Welding Institute, UK). In, *Proc. 4th Eur. Electro-Optics Conf., Society of Photo-Optical Instrumentation Engineers*, Vol. 164, pp. 279-287. Society of Photo-Optical Instrumentation Engineers, Bellingham, WA, USA, 1979.

The relevant laser process and material parameters are discussed in terms of their effect on welding performance and compared with other welding methods with particular attention being paid to weld depth.

4.1.2.1 (13) Aspects of using a laser for welding thick metal

Fedorov, V. G. et al. *Izv. V.U.Z. Mashinostroenie*, (8):115-118, 1980. (In Russian)

Description of standard procedures for laser welding of sheets of thickness exceeding 6mm, showing that requirements on the accuracy of weld preparation could be reduced by feeding additional materials to the welding zone and that a good fusion of the parent and the deposited metal could be ensured by scanning the beam. (10 refs.)

4.1.2.1 (14) Manufacturing methods and technology application of high energy laser welding process

Melonas, J. V. (Army Missile Command, USA). Final Tech. Rept. No. AD-A158082/8/XAD, 143pp., August, 1980

A 50kW CW CO_2 laser was used to establish welding parameters for common joint configurations for both steel and aluminium alloys; and analyse economic factors which prevail in a production environment. Feed rate, heat ranges, special tooling, and economic considerations were evolved for each metal. A significant breakthrough in the control of weld heat in the aluminium workpiece was evolved with the design of a reflecting shield which essentially increased the heat in the workpiece by a factor of two. Thus, greater depth and quality of the weld bead resulted.

4.1.2.1 (15) Laser welding of metals

Grigoryants, A. G. *Svar. Proizvod.*, 28(12): 14-16, 1981. (In Russian)

Studies CO_2 laser welding parameters and mechanical properties of welds in 12Kh2N4A, 18KhGT, Kh15N5D2T and grade 35 steels and Ti alloys VT3 and VT7. Tensile strength of experimental welds was comparable to those of welds produced by other processes. Fine-grained structure of the HAZ resulted in improved resistance to cracking. Welding speeds were 3.5×10^{-2} and 5×10^{-2} m/s for steels and Ti alloys, respectively. (8 refs.)

4.1.2.1 (16) Laser welding: state-of-the-art-review

Mazumder, J. In, *Proc. Lasers in Metallurgy Conf.*, 22-26 February 1981, Chicago, IL, pp. 221-245. Metallurgical Society/AIME, Warrendale, PA, USA, 1981. Also *J. Metals*, 34(7): 16-24, 1982.

The laser's capability for generating high power densities is a primary factor in establishing its potential for welding. Numerous experiments have shown that the laser permits precision weld joints of a high quality only rivalled by an electron beam. Recent work on laser welding of steels, Ni, Ti6Al4V and Al alloys 2219 and 5456 is reviewed. (71 refs.)

4.1.2.1 (17) Welding with high-power lasers

Banas, C. M. and Breinan, E. M. In, *Advances in Metal Processing, 25th Sagamore Army Materials Research Conf.*, 17-21 July 1978, Lake George, NY, pp. 111-131. Plenum Press, NY, 1981.

The performance and capabilities of laser welding are discussed. The advent of multi-kilowatt continuous wave laser systems has made possible the welding of heavy sections. Successful welds have been achieved in HY-130 and X-80 alloy steels of thicknesses > 1in. Ti, Al and stainless steel have all been welded with laser techniques. Laser welding has a high ratio of weld depth-to-width, and an ability to produce fusion zone purification by the actual removal of non-metallic inclusions. Weld penetration, shown as a function of laser power in kilowatts, is related to total beam power and focused power density. Travel speeds of 2-5ft/min have been obtained when welding 0.5in. thick Grade B ship steel. (35 refs.)

4.1.2.1 (18) Formation of the surface of the weld spot in laser welding

Davydov, V. A. and Magnitskii, O. A. *Weld. Prod.*, 29(12): 16-17, 1982.

It has been found that the formation of the spot surface in laser welding is strongly affected by the gas medium. In welding the majority of the examined metals, the use of shielding gases and gas mixtures results in smoothing the surface of the weld spot and prevents formation of internal defects.

4.1.2.1 (19) Laser welding of P/M materials

Lampugnani, U. et al. In, *P/M-82 in Europe, Int. Mettalurgy Conf.*, 20-25 June 1982, Florence, Italy, pp. 193-200. Associacion Italiana di Metallurgia, Milan, Italy, 1982. (Shorter version of paper in *Powder Metall., Int.*, (3): 115-118, 1983.)

The peculiarities and the advantages of laser interaction with porous structures were examined. Unlike standard welding processes, laser welding shows that results depend not only on the materials' compositions but also on sintering conditions. Low-carbon (C10) sintered steels can be satisfactorily welded provided that their structure is clean enough. Medium-carbon compositions can also give acceptable welds but preheating is necessary. Up to now some materials like Al 201AB alloy cannot be considered laser weldable; it seems impossible to avoid blowholes formation.

4.1.2.1 (20) Laser welding of steel and non-ferrous metals with a 200W Nd:YAG laser

Chen, G. and Ruge, J. *Metall*, 36(4): 409-413, 1982. (In German)

The absorption of laser energy, depends on the material to be welded, and can be increased by a suitable surface treatment. Penetration and volume of fused and evaporated metal were ascertained, taking into account the pulsed laser energy. (25 refs.)

4.1.2.1 (21) High-energy beams break into the big time

Mayer, C. A. *Weld. Design Fabric.*, 56(8): 48-55, 1983.

Electron beam and laser beam welding focus high-energy beams to concentrate power on the surface of material to be joined or cut. Depth-to-width ratios for butt welds can run as high as 10:1 for LBW, and single-pass butt welds without fillers may be ¾in. thick with laser welding. Compared to other processes, beam welding gives less distortion because of lower heat input and can save on welding time, filler metal and post-weld finishing. Lasers use a variety of light (photon) sources genererating heat and a cooling system continuously removing heat from the laser. Examples of beam welding are automobile clutch housings of 1010 steel and Inconel 625 recuperators for AGT 1500 turbine engines.

4.1.2.1 (22) Kawasaki Steel uses laser welder to join sheets

Jpn. Met. Bulletin, (4434): 1-2, 1983.

A fully automated CO_2 laser welder developed to join coiled sheet lengths previous to rolling or pickling has been installed in combination with a high-precision mechanical shear tool. Sheets up to 6mm thick and 1900mm wide can be welded, at a rate of 5m/min on 2mm thicknesses.

4.1.2.1 (23) Laser welding for hermetic sealing of metallic miniature housing

Daene, K. et al. (Zentralinstitut fuer Schweisstechnik der DDR, East Germany). *Feingeraetetechnik*, 32(12): 556-560, 1983. (In German)

Solid-state laser welding installations using Nd:YAG lasers have proved to be especially suited for the welding of miniature parts. High-quality spot welds can be produced with a laser pulse, whilst hermetically sealed weld seams can be produced with pulse sequences. The laser beam is characterised by a high power density, contactless operation with a light beam, no electric or material contact with the heat source, low tool wear, and high reproducibility of the operational and quality parameters.

4.1.2.1 (24) Increasing the efficiency of penetration in laser welding with dynamic beam focusing

Ivanov, V. V. *Weld. Prod.*, 31(5): 17-18, 1984.

Penetration depth of steels, titanium, and aluminium alloys can be increased by 30-45%, with a reduction in variation of depth of penetration, an improvement in the stability of weld surface formation, and an increase in the total thermal efficiency.

4.1.2.1 (25) Narrow seam laser welding of thin components

Levin, G. I. *Weld. Prod.*, 31(5): 22-23, 1984.

Transfer of metal in the vapour phase is a feature of the mechanism of seam welding small thicknesses with deep penetration for which the pulsed CO_2 laser is suitable and, at present the only tool. The radiation parameters of this laser coincide with the parameters required to ensure minimum weld formation.

4.1.2.1 (26) Welding of thick sheet metals by means of carbon dioxide lasers

Wolf, G. *Maschinenmarkt.*, 90(103/4): 2477-2478, 1984. (In German)

Laser beam welding has several advantages over other welding methods: high energy densities of the laser beam, no electrode consumption, high welding speeds and no need for specific welding atmospheres. From more than 50 different kinds of lasers, however, only the CO_2 laser is at present successfully applied for the welding of thick sheet metals. Reviews CO_2 laser welding technology and the properties of a laser beam which depend on the laser set-up and the configuration of the resonator. For the focusing and aligning of laser beams, several different lenses and mirrors are available.

4.1.2.1 (27) Welding technology setting high standards

Kubel, E. J. *Materials Engng.*, 99(5): 39-46, 1984.

Mechanical properties of laser welds in X-80, HY130, X-60, HY80 and BS 1501-271 Gr-W30 steels are listed. Electron beam and laser welding for mechanically strengthened oxide dispersion strengthened alloys are also presented.

4.1.2.1 (28) Laser welding of metal. III – The role of assist gas in CO_2 laser welding

Arata, Y., Maruo, H. and Miyamoto, I. *Q. J. Jpn. Weld. Soc.*, 3(2): 276-283, 1985. (In Japanese)

The results obtained are summarised as follows:- when the assist gas pressure p is higher than p_p, which is slightly higher than vapour pressure or surface tension pressure, the plasma formation is suppressed by forcing the vapour stream going up along the laser beam axis to eject along the rear wall of the cavity providing increased penetration without weld defects. The value p_p decreases with increasing laser power and decreasing welding speed. The fluctuation of the penetration tends to increase with increasing p, since the cavity diameter is enlarged by the vapour pressure which exceeds the value for the surface tension pressure to keep the cavity stable. When p is too high, a lot of molten metal flows out of the cavity closing the cavity entrance on re-entering into the cavity due to its large mass motion, providing a humping bead containing large porosities. The plasma formation is also suppressed effectively at vacuum pressure range 10-20 Torr, providing deeper penetration depth without fluctuation of penetration. (15 refs.)

4.1.2.1 (29) The development of the 'Laspot' welding process

Bazan, M. et al. (Ohio State University, USA). In, *Proc. 3rd Int. Conf. on Lasers in Manufacturing*, 3-5 June 1986, Paris, France, (Ed. A. Quenzer), pp. 107-116. IFS (Publications) Ltd, Bedford, UK, 1986.

A patterned laser spot welding process was developed as an alternative to resistance spot welding. These laser spot welds are visually inspectable and are at least as strong as resistance spot welds. A prototype welding system consisting of a carbon dioxide laser, a computer numerical controlled positioning table, and a clamping device was developed for the program. Laser spot welds were produced by programming the computer numerical controller to move the positioning table in a desired pattern under a stationary focused laser beam. The three patterns programmed into the controller were a spiral, a centreless spiral, and a circle. The laser spot welds produced have an outer diameter equal to, or less than the resistance spot weld indentation diameter. Laser spot welds of the various weld configurations and optimum resistance spot welds were produced in Hastelloy X, 321 stainless steel, and Inconel 718. Room temperature tensile shear tests were the basis of strength comparison.

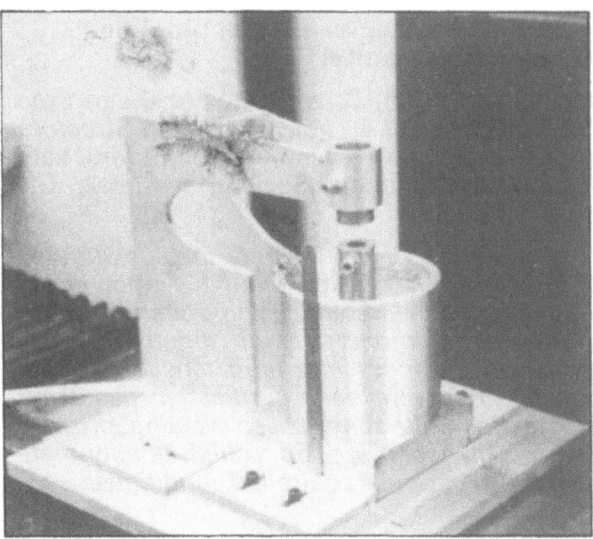

The prototype clamping device for the Laspot welding apparatus (See 4.1.2.1 (29))

4.1.2.2 Laser welding of ferrous metals

4.1.2.2 (1) Significance of variables

Burris, K. (Bendix Corp, USA). SME Tech. Paper No. AD75-872, 13pp. Bendix Corp, Kansas City, MO, USA. Presented at Advanced Welding and Brazing Workshop, 28-30 October 1975, Chicago, IL.

Studies have been conducted to determine the significance of process variables in making continuous laser welds in stainless steel parts for precision electromechanical equipment. Laser beam power and weld speed are the principal determinants of weld nugget width, depth, and microstructure; and the flow rate of the cover gas CO_2 determines the post-weld surface condition of the weld area.

4.1.2.2 (2) Establishment of a continuous wave laser welding process

Seaman, F. D. (Sciaky Bros. Inc, USA). Final Rept. No. AD-B022 759/5/XAD, 338pp., October 1976. Sciaky Bros. Inc, Chicago, IL, USA.

Determines the maximum single pass thickness that could be welded using a 15kW CO_2 laser. Also laser welded test plates in representative aircraft materials were prepared, subjected to non-destructive examination and mechanically tested to failure. Laser welds generally equalled base metal performance in fatigue and tensile tests. The fracture toughness of laser welds in

high strength steel was shown to be very high, particularly when tests were conducted in the presence of synthetic sea water.

4.1.2.2 (3) Evaluation of basic laser welding capabilities

Breinan, E. M. et al. (United Technologies Research Center, USA). Tech. Rept. No. UTRC/R77-911989-10, 59pp, March 1977. United Technologies Research Center, East Hartford, CT, USA.

Hy-130 alloy steel was laser welded in a series of thicknesses including 0.64, 0.95, and 1.27cm at continuous laser powers ranging from 6.0 to 12.8kW and speeds between 1.27 and 2.96cm/sec. All welds were visually inspected, X-rayed, magnafluxed, and subjected to a series of mechanical tests including hardness, bend, tensile, impact, and dynamic tear tests. Although some of the specimens, primarily those of intermediate thickness (0.95cm), encountered porosity problems during laser welding, all specimens performed extremely well in mechanical tests.

4.1.2.2 (4) Laser welding at 100kW

Laser Focus, 13(3): 14, 16, 18, 20, 1977.

Describes tests performed with a "triservice" laser. The laser amplified a fundamental mode, low-power input beam. With multiple passes of the laser cavity, peak output can reach 100kW for a few seconds and weld 2 inch steel plates at 50in/min. A discussion is outlined of the potential problems of industrial applications of lasers.

4.1.2.2 (5) Laser welding of pipeline steels (initial results)

Banas, C., Parrini, C. and Vito, A. D. In, *Proc. Welding of HSLA (Microalloyed) Structural Steels Conf.*, 9-12 November 1976, Rome, Italy, pp. 604-624. American Society for Metals, OH, USA, 1978.

Examination of laser welds in pipeline steel (X60 and X65) revealed relatively poor visual, metallographic and radiographic characteristics. High values of impact strength, together with the increased hardness and high tensile strength, indicate that some zone purification has occurred in the weld zone. Laser welding provides best results when applied to high-quality materials under carefully controlled conditions.

4.1.2.2 (6) Arc augmented laser welding

Steen, W. M. and Eboo, M. (Imperial College, UK). *Met. Constr.*, 11(7): 332-333, 335, 1979. (Also in *Proc. Advances in Welding Processes Conf.*, Vol. 1, 9-11 May 1978, Harrogate, UK, pp. 257-265. Welding Institute, Cambridge, UK, 1978.)

A study of the combination of an arc welding and a laser welding process to assess the advantages. A CO_2 laser and a water-cooled TIG welding set were mounted either above or below a substrate. When the arc was placed on the same side of a plate as the laser, a 50% increase in welding speed for a low current was obtained. With the arc on the opposite side of a 0.2mm thick mild steel sheet, the welding speed increased fourfold. Other advantages include a narrow root giving a more intense action and the absence of undercutting associated with high speed welding processes. (15 refs.)

4.1.2.2 (7) The effect of roughness of the metal on the depth of melting in pulsed laser welding

Bashenko, V. V. et al. *Avtom. Svarka.*, (5): 41-42, 1979. (In Russian)

The effect of micro-roughness was studied with specimens of stainless steel 12Kh18N10T. The depth of melting increased with a reduction in the height of micro-roughness.

4.1.2.2 (8) The laser welding of gear wheels made of steel 40Kh

Alekseev, V. A. et al. *Svar. Proizvod.*, 26(9): 17-18, 1979.

A technology is proposed for the laser welding of steel gear wheels. The results of mechanical tests and metallographic examination are briefly given.

4.1.2.2 (9) Laser welding of steel 12Kh18N10T

Avramchenko, P. F. et al. *Autom. Weld.*, 33(3): 32-33, 1980. (See also paper by some authors in *Autom. Weld*, 33(8), 1980 on welding thin-walled pipes of the same steel.)

Laser butt welds were obtained on stainless steel 12Kh18N10T of 5-6mm thickness at 5-23kW. The gap between the sheet assembled for welding was 0.1mm. Satisfactory welds were obtained. The use of fluxes (AN-60) and helium protection was suggested to prevent weld metal oxidation.

4.1.2.2 (10) Applicability of laser welding to steel strip for cold rolling

Nishiyama, N., Sasaki, H. and Tsuboi, J. *Kawasaki Steel Giho*, 13(3): 423-430, 1981.

Discusses the applicability of CO_2 laser welding to hot rolled steel coils with emphasis on determining the best welding conditions to suit each production line, and on increasing fit-up tolerance. Welds in various steels were tested for rollability using reverse bend tests and miniature mill rolling. The weld joints in 3% Si steels showed good rollability and it was concluded that CO_2 laser welding was satisfactory.

4.1.2.2 (11) Comparison of laser welds in two stainless steels

Jones, T. A. (Oak Ridge Y-12 Plant, USA). Rept. No. Y/DV-180, 45pp., November 1981.

304L and 316 stainless steel welds were investigated and there were no significant differences in the weld bead characteristics. When the pulse length is increased, the spot size and spot overlap decreases, but the penetration remains the same. Increases in frequency result in an increase in percent overlap and decrease in spot size and penetration. When the power is increased, the penetration, overlap, and spot size increase.

4.1.2.2 (12) High-power laser beam welding of T-sections with different edge preparations

Hakansson, K. In, *Proc. JOM-1 (Joining of Metals) Conf.*, 9-12 August 1981, Helsingor, Denmark, pp. 36-43. Ingeniorhojskolen Helsingor Teknikum, Helsingor, Denmark, 1981.

Discusses the high-power laser beam welding of T-sections made of 16mm ship steel. Five different edge preparations were studied, to determine if the round rolled edges of the web could be satisfactorily welded to the flat side of the flange. Filler material fed into the joint during welding served to fill the gap in bevelled specimens and to establish a fillet in square joint specimens.

4.1.2.2 (13) Laser welding of high-strength steels used in transportation industry

Blarasin, A. (Fabbrica Italiana Automobil, Italy). *Riv. Ital. Saldatura*, 33(5): 267-281, 1981 and 33(6): 352-360, 1981. (In Italian)

A 15kW AVCO continuous wave CO_2 laser has been used for evaluating the influence of shielding gas, focal point position, welding speed and power on depth of penetration. Tests have been carried out on two Nb microalloyed HSLA steel sheets, 2mm and 8mm thick. Laser welded joints showed high fatigue resistance. Metallographic details for the laser welds are given, and the maximum allowable gap between plate edges in laser butt welder joint is evaluated. (13 refs.)

4.1.2.2 (14) Laser welding of steel plates with unmachined edges

Hill, M., Johnson, R. and Megaw, J. H. C. In, *Proc. The Joining of Metals: Practice and Performance Conf.*, Vol. 2, 10-12 April 1981, Coventry, UK, pp. 146-154. Institution of Metallurgists, London, UK, 1981.

Satisfactory cutting performance was achieved in these trials when using air as the assist gas. Although the alternative use of high pressure oxygen yielded very significant enhancement of cutting speed, air has the merit of economy and of yielding a superior edge finish. Helium-cut edges were much less oxidised than air-cut edges but, when welded without filler, tended to exhibit weld porosity. With the present laser welding technique, the grade 43A parent material may exhibit small solidification cracks. This appears to correlate with levels of S, P and C. The use of automatic wire feed apparatus in conjunction with laser welding is controllable and reproducible, and double-deoxidised filler is found to solve all problems associated with parent plate composition, cut edge condition and plate surface condition.

4.1.2.2 (15) The laser welding of steels used in can making

Mazumder, J. and Steen, W. M. *Weld. J.*, 60(6): 19-25, 1981.

Sound welds of good appearance and mechanical properties can be made using a laser on tinplate and Sn-free steel. The welding speeds are low compared to present can making speeds; however, speed can be increased by

using an electric arc. Laser welds had a very narrow HAZ and could be made through painted areas. The laser is possibly the only known method of welding or joining Sn-free steel at atmospheric pressure without auxiliary preparation. (17 refs.)

4.1.2.2 (16) A study on in-process assessment of joint efficiency in the laser welding process

Hoshinouchi, S. et al. In, *Proc. Fundamental and Practical Approaches to the Reliability of Welded Structures Conf.*, Vol. 1, 24-26 November 1982, Osaka, Japan, pp. 175-180. Japan Welding Society, Tokyo, Japan, 1982.

A monitoring technique for inspection and evaluation of penetration in laser welding was developed. The technique is based on the phenomemon that the metal irradiated laser beam emits a specific sound. Some welding experiments (bead-on-plate) on stainless steel plates were tried to pick up the weld sound. It was shown that penetration can be monitored by measuring the sound pressure level.

4.1.2.2 (17) Welding of vacuum arc melted 03Kh11N10M2T steel by the continuous CO_2 laser beam

Belen'kii, A. M. et al. *Weld. Prod.*, 29(1): 16-18, 1982.

Thicknesses up to 5mm were welded. Satisfactory formation of single-pass welds is ensured within a wide range of welding conditions: welding speed 60-80m/h, beam power 1.3kW/1mm of thickness. The width of the zone of softening in the laser welds is a factor of 3-4 smaller than in Argon arc welding and the level of welding strains is a factor of 8-10 lower. The corrosion resistance of the laser welded joints is satisfactory.

4.1.2.2 (18) Advances in laser and MIAB welding techniques

Crafer, R. C., Edson, D. A. and Johnson, K. I. *Weld. J.* 62(2): 15-20, 1983.

Laser welding is a new technique which shows increasing potential for use in the mass production and aerospace industries for welding components up to 13mm thick. The technique is fast, relatively clean and eliminates production and design constraints imposed by other processes. Recent work conducted to improve welding capability and equipment reliability is described and the results of recent welding trials using mild, low-alloy and stainless steels are presented.

4.1.2.2 (19) Effect of the composition of shielding gases on weld formation in laser welding

Kovalev, V. V., Surkov, A. V. and Novozhilov, N. M. (Central Scientific Research Institute of Heavy Engineering, USSR). *Weld. Prod.*, 30(9): 27-28, 1983.

The dimensions of the joints depend on the composition of the welded material and change in a non-linear manner with the variation of the composition of the shielding gas. Presents results of investigations into the effectiveness of penetration in laser welding of a pearlitic and an austenitic steel in various shielding gases.

4.1.2.2 (20) Laser welding of mild steel at NIROP, Minneapolis, USA

Metzbower, E. A. *Met. Constr.*, 15(10): 611-613, 1983. (Also Naval Research Laboratory Rept. No. NRL-8646, 22pp., February 1983).

A description is given of the continuous wave CO_2 laser welding system at the Naval Industrial Reserve Ordnance Plant, Minneapolis, Minn., for the fabrication of guided missile launching systems and gun mounts from A36 steel (0.29% carbon (max.), 0.8-1.2% Mn, 0.15-0.45% Si). Details are given of the laser output, control system and focusing. Process parameters, weld strength, toughness and microstructure are reported for trials on the steel. Economic advantages of the system are discussed. Practical experience is described and cost savings compared with other welding processes are estimated.

4.1.2.2 (21) Special features of laser welding of medium-carbon steel through a cadmium coating

Ofer, V. I. and Uglov, A. A. *Phys. Chem. Mater. Treat.*, 17(4): 400-405, 1983.

Experimental data are presented on the effect of a Cd coating on the formation of melting down zones during laser welding of medium carbon steel. A significant increase in the thermal effect zone was observed with an increase of thickness of the coating and the density of the radiation flux.

4.1.2.2 (22) Development of laser welder for strip processing line

Kawai, Y. et al. (Kawasaki Steel Corp, Japan). Kawasaki Steel Tech. Rept. No. 10, pp. 39-46, December 1984.

Kawasaki Steel Corp. has developed CO_2 laser welders for the sheet steel production process. The 5kW CO_2 laser welders installed for pickling lines at Mizushima and Chiba Works have unst-

able resonators and wire feeders. These lasers can weld all kinds of steel strips measuring up to 6mm in thickness, and up to 1880mm in width.

4.1.2.2 (23) An introduction to CO_2 laser welding low-carbon steel up to 4mm thick

Dawes, C. J. In, *Proc. Developments and Innovations for Improved Welding Production 1st Int. Conf.*, 13-15 September 1983, Birmingham, UK, pp. 43.1-43.10. Welding Institute, Cambridge, UK, 1984.

The laser, workhandling equipment, and focusing optics are briefly described. Several sheet metal joint configurations which are suitable for laser welding are illustrated and general laser welding process data are presented. The difference in laser power and welding speed combinations for the production of acceptable welds in different joint configurations is illustrated by reference to butt and T joints. Workpiece fit-up information shows that only very small gaps can be tolerated when making autogenous welds – approximately 0.12mm in 2mm sheet butt joints – whereas satisfactory butt welded joints can be made with a 50% sheet surface mismatch. Gas shielding is briefly described and shows that helium is currently the preferred shielding gas. (12 refs.)

4.1.2.2 (24) New techniques for welding sheet and tubular fabrications

Dawes, C. J., Edson, D. A. and Johnson, K. I. In, *Proc. Sheet Metal Welding Conf.*, 30 October – 1 November 1984. Detroit, MI, Paper 9, 23pp. American Welding Society, Miami FL, USA, 1984.

Describes the status of application of high power CO_2 lasers together with the results of recent work conducted to establish the tolerance to various process and fit-up parameters when laser welding butt and T joints in 2, 3 and 4mm. thick low-carbon steel sheets. Results are also presented on the advantages of using a recently developed laser beam spinning technique which gives increased tolerance to workpiece fit-up. The techniques also allowed the welding of material with guillotined edges and allowed substantial relaxation in laser beam/workpiece alignment. (2 refs.)

4.1.2.2 (25) Studies of combination type CO_2 laser welding system

In, *Proc. Quality and Reliability in Welding Conf.*, Vol. 1, 6-8 Sept. 1984, Hangzhou, China, 6pp. Welding Institution of the Chinese Mechanical Engineering Society, Harbin, China, 1984.

Investigates 500W grade combination CO_2 laser welding system made up of several sepa-

rate lower power lasers. Some factors which influence the properties of focused combination laser beam are analysed. By this system, miniature thin-wall stainless steel parts can be welded precisely. The advantages of the system are long-term stability and high quality of weld, compact size and ease of fabrication.

4.1.2.2 (26) Absorption of CO_2 laser beam by AISI 4340 steel

Khan, P. A. A. and Debroy, T. (Pennsylvania State University, USA). *Metall. Trans. B.*, 16B(4): 853-856, 1985.

Describes a technique to determine absorptivity under conditions of laser welding involving measurements and calculations using transient temperature profiles in the solid region of the weld. The absorptivity obtained by this technique is in agreement with that determined previously and with the theoretical value estimated using electrical resistivity data.

4.1.2.2 (27) Application of laser beam welding to thin steel sheet

Ishiyama, H., Masumoto, I. and Shinoda, T. *Jpn. Weld. Soc. Q. J.*, 3(1): 26-31, 1985. (In Japanese)

This experiment was carried out to find what advantage can be expected by applying laser beam welding to surface coated steel sheet and austenitic stainless steel. The results obtained and compared with TIG welding are as follows: steel sheet can be welded with much less distortion by laser beam welding. For coated sheet, laser beam welding causes less damage to coated layer. It is possible to make a full penetration weld 3-5 times faster by laser beam welding. TIG weld metal of Zn coated sheet shows Zn rich region at surface but laser weld metal does not, and laser beam welded joints of stainlees steel show little weld decay tendency. (7 refs.)

4.1.2.2 (28) Non-destructive control of weldings using the mirage detection

Boccara, A. C., Fournier, D. and Lepoutre, F. *J. Appl. Phys.*, 57(4): 1009-1015, 1985.

Small defects in edge-to-edge laser weldings of thin pieces of stainless steel have been detected. A simple model shows the physical parameters influencing the experimental results. Using the phase of both the transverse and normal deviations at various frequencies allows for the separation of optical and thermal effects, of surface and subsurface defects, and for the determination of the depths and widths of the subsurface defects. (14 refs.)

4.1.2.2 (29) Shielding gas effects in pulsed carbon dioxide laser spot welding

Chennat, J. C. (Photon Sources Inc, USA) and Albright, C. E. (Ohio State University, USA). In, *ICALEO '84, Proc. Materials Processing Symp.*, Vol. 44, 12-15 November 1984, Boston, MA, (Ed. J. Mazumder), pp. 76-85. Laser Institute of America, Toledo, OH, USA, 1985.

The effect of shielding gases and laser variables on weld fusion zone configuration have been investigated in the pulsed spot welding of AISI 1018 steel using a CO_2 laser with 610W output power. Laser parameters chosen were tube discharge current levels of 50, 100, 150mA; laser pulse lengths of 0.5, 1, 2, 4, 8, and 16ms and focal positions of 2.5mm. above, at, and 2.5mm. below target surface. He, CO_2, N_2, Ar as individual gases and their binary gas mixtures at 50cfh and air (no flow) were the shielding gases studies. Average laser power densities in the range of 0.3 to $0.8 \times 10^6 W/cm^2$ were employed.

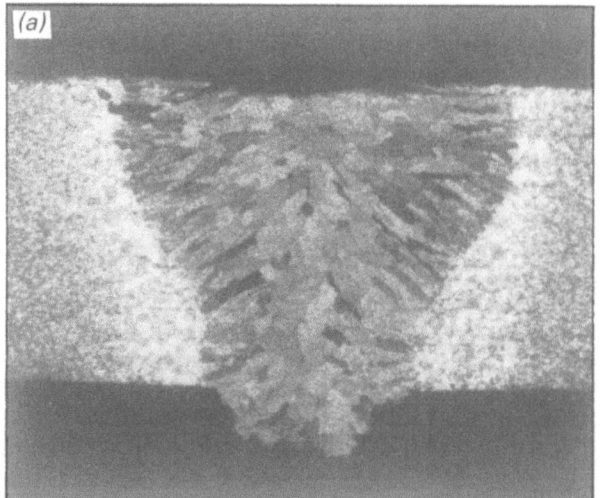

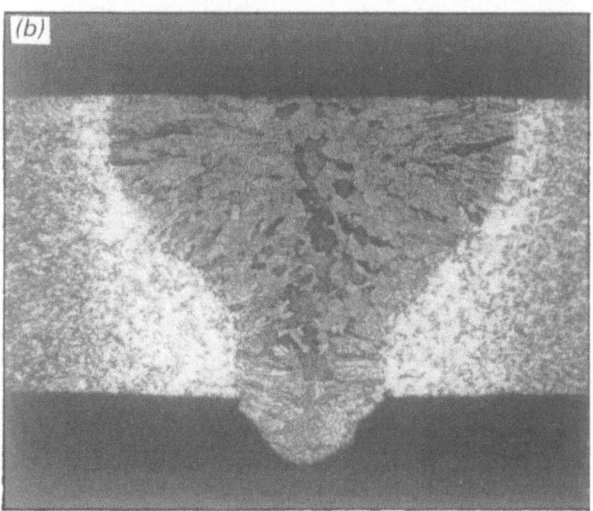

Comparison of: (a) 450W laser + 180W auxiliary source, with (b) 1kW laser (See 4.1.2.2 (31))

The fusion zone configuration studied included weld penetration, weld pool diameter and fusion zone volume. Spectrographic analysis of the laser plasma plume was undertaken to identify the predominant species in the plasma plume with He and Ar gas shielding.

Results show that the depth of penetration is determined by discharge current level and focal position. Laser pulse length is found to influence the fusion zone volume. With regard to their beneficial effect on improved weld penetration and fusion volume, the shielding gases listed in order of maximum to minimum benefit are He, CO_2, air, N_2, and Ar. *(3 refs.)*

4.1.2.2 (30) Study of the quality of butt joints in sheets as a function of weld seam variations

Ganyushin, V. M. et al. *Avtom. Svarka.*, (1): 37-41, 1985. (In Russian)

The effect of sheet metal thickness, welding process and welding conditions on quality of butt welds made in 012Kh18N10T steel was studied. Bronze was found to be the best sheet clamping material. Geometric parameters of the welded butt seams exhibited a linearly increased deviation as sheet metal thickness increased from 0.12 to 0.30mm. Laser welding was used successfully for the thinnest sheet. *(10 refs.)*

4.1.2.2 (31) CO_2 laser auxiliary source coupling: application to welding

Luciani, P. Y., Charissoux, C. and Calvet, J. N. (Commissariat à l'Energie Atomique (CEA), Centre d'Etudes Nucléaires de Saclay, France). In, *Proc. 3rd Int. Conf. on Lasers in Manufacturing*, 3-5 June 1986, Paris, France, (Ed. A. Quenzer), pp. 117-124. IFS (Publications) Ltd, Bedford, UK, 1986.

By preheating the material, the coupling of the CO_2 laser with an auxiliary source serves to reduce the reflection coefficient and hence increase laser penetration. The system was characterised on austenitic stainless steel plates (AISI 304L) and other reflective materials. The experiments show that for a laser power of 450W and an auxiliary source power of 180W, penetration is equivalent to that obtained with a 1kW laser. Within the framework of process automation, this type of coupling system allows accurate positioning of the plate in relation to the lens focal zone, by using the arc voltage regulation of the auxiliary source.

4.1.2.2 (32) Examples of possibilities offered by high power lasers in welding applications

Bousseau M. and Signamarcheix J.- M. (Estab-

lissement Technique Central de l'Armement (ETCA), France). In, *Proc. 3rd Int. Conf. on Lasers in Manufacturing*, 3-5 June 1986, Paris, France, (Ed. A. Quenzer), pp. 125-134, IFS (Publications) Ltd, Bedford, UK, 1986.

Beyond the parameters of power and speed, laser welding depends on focus spot and energy distribution. The used protection gas interacts directly on plasma expansion and limits the beam penetration. With ETCA's 5kW laser, welding depths of 8mm. are obtained in the case of steels.

In this paper, results on steels, which are used to build capacities under pressure (type 15CDV6), stainless steels and high strength steels are given. Micro-structural transformations and mechanical properties induced by welding are analysed. The point welding laser to body steels is also investigated.

4.1.2.2 (33) Laser welding and its applications for steel making process

Shinmi, A. et al. (Mitsubishi Electric Corporation and Manufacturing Development Laboratory, Japan). In, *Laser Welding, Machining and Materials Processing – Proc. Int. Conf. on Applications of Lasers and Electro-Optics, ICALEO '85*, 11-14 November 1985, San Francisco, CA, (Ed. C. Albright), pp.65-72. Laser Institute of America, Toledo, OH, USA/IFS (Publications), Ltd, Bedford, UK, 1986.

Investigates welding characteristics of annular beam mode CO_2 laser which is mostly used in steel process as a CO_2 laser welder. Furthermore, optimum butt-welding conditions of shear-cut medium thick steel sheet (less than 10mm thick) and thin steel sheet which is cut by several methods including shear-cut (more than 0.15mm thick) are described. Finally, as a typical application, 5kW CO_2 laser welder, which was employed in the existing continuous pickling line, is introduced.

4.1.2.2 (34) Laser welding of sheet metal fabrications – process improvements

Dawes, C. J. (The Welding Institute, UK). In, *Laser Welding Machining and Materials Processing – Proc. Int. Conf. on Applications of Lasers and Electro-Optics, ICALEO '85*, 11-14 November 1985, San Francisco, CA, (Ed. C. Albright), pp. 73-80. Laser Institute of America, Toledo, OH, USA/IFS (Publications) Ltd, Bedford, UK, 1986.

Laser welding research on sheet metal joints, in low-carbon deep drawing and HSLA steels up to 4mm thick, has shown that joint fit-up and laser beam alignment tolerances can be considerably increased by employing beam spinning. *(6 refs.)*

General view of laser laboratory at ETCA (See 4.1.2.2 (32))

4.1.2.3 Laser welding of non-ferrous metals

4.1.2.3 (1) Laser welding of titanium 6AI-4V

Mazumder, J. and Steen, W. M. (Imperial College, UK). In, *Proc. Laser 77: 3rd Optic-Electronics Conf.*, 20-24 June 1977, Munich, W. Germany, pp. 307-315. IPC Science and Technology Press, Guildford, UK, 1977.

Describes the relationship between the main welding parameters such as laser power, welding speed, substrate thickness on fusion width, HAZ and weld properties. The mechanical properties of the CO_2 laser welds are evaluated together with their corrosion resistance. The structure of the weld zone is examined and the observed structures compared to those expected for the abrupt temperature cycle of a laser weld. (15 refs.)

4.1.2.3 (2) Welding uranium with a multi-kilowatt, continuous wave, carbon dioxide laser welder

Turner, P. W. and Townsend, A. B. (Oak Ridge Y-12 Plant, USA). Rept. No. Y-2079, 23pp, June 1977.

A 15kW CO_2 laser made partial penetration welds in 6.35 and 12.7mm. thick wrought depleted uranium plates. Welding power and speed ranged from 2.3 to 12.9kW and from 21 to 127mm/s. Results show that depth-to-width ratios of at least unity are possible and the overall characteristics of the process indicate it can produce welds resembling those made by the electro beam welding process.

4.1.2.3 (3) Evaluation of basic laser welding capabilities

Snow, D. B. and Breinan, E. M. (United Technologies Research Center, USA). Tech Rept. No.

UTRC-R78-911989-14, 41pp, July 1978. United Technologies Research Center, East Hartford, CT, USA.

5456 and 5086 Aluminium alloys were welded at 1.2-7.2cm/s. with a continuous CO_2 laser operated at beam powers of 5-9kW. Difficulty was experienced in reproducing weld bead geometries and depths of penetration from weld to weld. Control of the helium shield gas flow rate and flow pattern were observed to be particularly critical for maintenance of full penetration of the fusion zone in both 0.64 and 0.95cm plates.

4.1.2.3 (4) Welding of copper silicon and zirconium II using the 1200W CO_2 continuous laser

Shaber, E. L. and Debban, B. L. (United Nuclear Industries Inc, USA). Rept. No. UNI-1190, 19pp; November 1978.

The investigation was conducted to find an improved and updated welding technique for coextrusion billet encapsulation components. Acquisition of a laser system is recommended. The system would initially be placed in a development laboratory for 'debugging', verification of safety considerations and training purposes.

4.1.2.3 (5) Application of laser welding to overcome joint asymmetry

Morgan-Warren, E. J. (CEGB, Marchwood Engineering Laboratory, UK). *Weld. J.,* 58(3): 76s-82s, 1979.

Experimental welds were made between titanium sheets, 1mm. and 12mm. thick, to simulate the tube-to-tube plate joint configuration. The penetration characteristics were greatly superior to those obtainable with conventional GTA welding. Laser welding is effective in overcoming the heat sink problems associated with joint asymmetry. Its advantages include a narrow weld bead and heat affected zone, reduced HAZ grain growth, reduced danger of atmospheric contamination and a high tolerance to variations in welding parameters and joint fit-up. The main problems to be overcome include weld bead root porosity.

4.1.2.3 (6) Evaluation of basic laser welding capabilities

Breinan, E. M. et al. (United Technologies Research Center, USA). Final Rept. No. UTRC/R79-911989-17, 44pp; July 1979. United Technologies Research Center, East Hartford, CT, USA.

5456 and 5086 aluminium alloys were welded

with a continuous, cross-beam CO_2 laser operated in the master oscillator/power amplifier mode. Extensive experimental parameter modifications were made to the welding apparatus in an attempt to improve weld quality, in particular the reproducibility and uniformity of fusion zone penetration. The effects of laser power, mode and optics, welding speed, beam rotation, shielding design, weld preparation, wire feed, flux and back-up parameters on weld performance were evaluated.

4.1.2.3 (7) Welding of Ti-6Al-4V by a continuous wave CO_2 laser

Mazumder, J. and Steen, W. M. (University of Illinois, USA). *Met. Constr.,* 12(9): 423-427, 1980.

The advantages of laser welding are outlined, and the relationship between the welding parameters found from a systematically prepared set of welds made using a Control Laser 2kW CW CO_2 laser and the Culham/Ferranti 5kW CL5 CW CO_2 laser is discussed. Considers welding variables such as laser power; welding speed; material thickness; and thermal, process and melting efficiencies. Welded joint microstructure and mechanical properties are evaluated.

4.1.2.3 (8) Laser welding beryllium in a deuterium atmosphere

Faulkner, G. E., Murchie, J. R. and Ramos, T. J. (Lawrence Livermore National Laboratory, USA). Rept. No. DE83003312, 17pp, November 1982.

The methods included direct fusion welds, braze welds and fusion welds in Al coatings on Be. Bead-on-plate laser welds were made in D, Ar and helium atmospheres. Severe porosity was observed in welds made in D atmospheres. Weld cracking was observed in welds made in inert gas atmospheres. Weld porosity was the major obstacle to laser fusion welding Be in D atmospheres. Braze welding with A1 reduced porosity, but the quality of braze welds was unacceptable. Fusion welds in high-purity Al coatings on the Be were sound. A hole in Be was successfully sealed by coating a counterbore around the hole with high-purity Al and laser welding the Al.

4.1.2.3 (9) Laser welding of a titanium alloy

Fraser, F. W. and Metzbower, E. A. In, *Proc. Seminar on Advanced Processing Methods for Titanium,* 13-15 October 1981, Louisville, KY, pp. 175-188. Metallurgical Society/AIME, Warrendale, PA, USA, 1982.

A series of autogeneous, laser beam weldments have been fabricated in 12mm. thick plates. These single-pass weldments were fabricated on a continuous wave CO_2 laser at a power level of 11kW and a speed of 14.8mm/s. Mechanical properties of the weldments were determined using tranverse tensile specimens. Charpy V notch and dynamic tear testing was carried out to measure fracture toughness at 0 and 25°C. Resistance of the weldments to stress corrosion cracking was determined. Hardness measurements were taken in the fusion zone, heat affected zone and in the unaffected base metal. Microstructures were identified in these three areas. (13 refs.)

4.1.2.3 (10) Applicability of thermal joining techniques for less common metals (Be, Ti, Zn, Hf, V, Nb, Ta, Cr, Mo, W)

Lison, R. (Kernforschungsanlage Juelich, Zentralabteilung Allgemeine Technologie, W. Germany). *Radex Rundsch*, 1(2): 99-117, 1983. (In German)

Apart from new processes, for example, diffusion welding, electron beam and laser welding, this article discusses variations in electric arc welding and soldering for non-ferrous metals. The required atmospheres for welding these special metals are discussed.

4.1.2.3 (11) Laser welding of aluminium and aluminium alloys

Eagar, T. W. and Huntington, C. A. *Weld. J.*, 62(4): 105-107, 1983.

Studies the effects of surface preparation and joint geometry on laser power absorption by pure Al and by Al alloy 5456. The results indicate that initial absorption varies from a few percent to >25%, depending on the surface preparation. The fraction of absorbed power increases dramatically on formation of a keyhole. As a result, welds made with sharp bevel-groove preparation are larger and more uniform than those made with either bead-on-plate or square-groove weld preparations.

4.1.2.3 (12) Special features of laser welding AMg6 alloy

Grigor'yants, A. G. et al. (N. E. Bauman Higher Technical School, USSR). *Weld. Prod.*, 30(9): 24-26, 1983.

Describes investigations into the weldability of Al alloys by the beam of a continuous CO_2 laser with special reference to manufacturing body components of larger sizes in which high strains can be created by ordinary arc welding procedures.

4.1.2.3 (13) Optical detection system for the evaluation of laser welds

Duncan, H. A. *Rev. Sci. Instruments*, 55(10): 1585-1589, 1984.

Quality control is a fundamental concern in laser welding operations. An optical technique in conjunction with signal processing and a microcomputer is used to obtain data from which the weld quality can be determined. The technique is one in which the plume that is generated on each impingement of the laser on the material (phosphor-bronze) is detected by a photodiode and processed.

4.1.2.3 (14) CO_2 laser welding of copper slabs

Dell'Erba, M. (Centro Laser, Italy). *Opt. & Laser Technol.*, 17(5): 261-262, 1985.

Briefly describes the laser welding of copper sheets up to 3mm thick with a 2kW CW CO_2 laser.

4.1.2.3 (15) Laser welder for fabricating neutron tubes

Borrego, J. H. and Barnaby, B. E. (Sandia National Laboratories, USA). Rept. No. SAND-84-1852, 86pp, March 1985. Sandia National Laboratories, Albuquerque, NM, USA.

A laser welder for use on very small vacuum devices fabricated from high-temperature melting point alloys and refractory metals has been developed at SNL. The welder includes a 400W, pulsed Nd:YAG laser, welding chamber, positioning system, closed-circuit TV monitor, and a computer controller.

4.1.2.3 (16) Laser welding of 6Al-4V titanium alloy

Jimenez, E. (Allied Corp, Bendix Kansas City Div., USA). Final Rept. No. BDX-613-3115, 30pp, May 1985, Bendix Corp, Kansas City, MO, USA.

The 6Al-4V-Ti welds were compared with gas tungsten arc and electron beam weld for porosity, ductility, and contamination. Test results demonstrate that laser welding is suitable for alloy and that sound welds can be made over a wide range of welding conditions.

4.1.2.3 (17) Laser welding of thermally sensitive alloy

Werth, D. L. *Lasers Appl.*, 4(3): 69-70, 1985.

Laser welding of various turbine components is discussed emphasising that lasers can produce results comparable to electron beam systems if proper attention is directed to the beam structure and other process parameters.

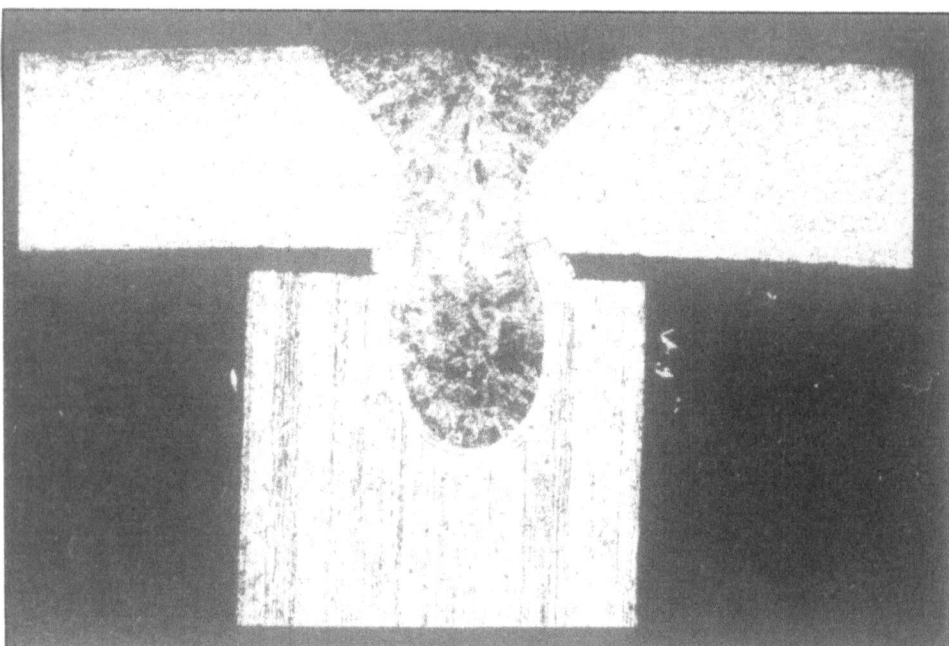

Cross-section of a 0.5/ 1mm T shape nickel weld (P=800W; v= 2.5m/min; foc = −0.5mm) (See 2.1.2.3 (18))

4.1.2.3 (18) Laser welding of thin nickel sheets

Petesch, B. et al. (CALFETMAT-IUT, France). In, *Proc. 3rd Int. Conf. on Lasers in Manufacturing,* 3-5 June 1986, Paris, France, (Ed. A. Quenzer), pp. 145-156. IFS (Publications) Ltd, Bedford, UK, 1986.

The laser welding of nickel sheets of 0.5mm and 1mm in thickness plate-to-plate and in T-shape by transparency are studied as follows. The welding parameters taken into account are the welding speed, the power and the focusing of the laser beam.
First, melting lines were done on stainless steel sheets (304 L, 7mm in thickness). Then, the joinings were achieved on nickel sheets (0.5 and 1mm in thickness). They are studied with metallorgraphic observations.

4.1.3 Laser welding of dissimilar metals

4.1.3 (1) Precision pulsed laser welding

Bolin, S. R. and Maloney, E. T. (Raytheon Co, USA). SME Tech. Paper MR75-571, 21pp, December 1975. Society of Manufacturing Engineers, Dearborn, MI, USA.

Applications are described as well as the material properties and their interacting relationships with the laser beam in successful fusion joining. Emphasises its non-contact nature, the minimal heat affected zone produced and the ability of the laser to spot or seam join certain material combinations better than by traditional techniques and its cost-effective nature.

4.1.3 (2) Advanced joining processes

Miller, F. R. (Air Force Materials Laboratory, USA). *SAMPE Q*, 8(1): 46-54, 1976.

Presents an overview of the Division's involvement in the development of a variety of advanced metals joining processes. Processes discussed include continuous wave laser welding.

4.1.3 (3) Investigations on laser welding

Montanarini, M. and Steffen, J. (LASAG, Switzerland). *IEEE J. Quantum Electron.,* QE12(2): 126-129, 1976.

A welding method is described that makes use of fundamental mode laser pulses which show single-mode relaxation oscillations with a high peak power first spike causing break-down of the material surface. The laser emission following the first spike can be adapted to the welding requirements and the material properties. Metallurgical problems of intermixing of dissimilar materials in the weld puddle have also been investigated. The lap weld of a stainless-steel spring with a nickel-iron baseplate has been performed with Nb as welding agent.

4.1.3 (4) Laser welding of dissimilar metals: titanium to nickel

Seretsky, J. and Ryba, E. R. (Pennsylvania State University, USA). *Weld. J.*, 55(7): 208-211, 1976.

Describes an unsuccessful attempt to carry out laser welding of titanium to nickel, which resulted in a cracked weldment which exhibits a poor microstructure. It was found that the cracking could not be eliminated by changing the laser power or by rewelding the samples.

4.1.3 (5) Laser welding of joints between tungsten alloy and stainless steel (EI395)

Voshchinskii, M. L. *Weld. Prod.*, 23(7): 27-28, 1976.

Investigates the possibility of laser welding dissimilar materials – steel Kh15N25M6 (15%Cr:25%Ni:6%Mo) and sintered tungsten alloy VNZh7-3 (W-Ni-Fe). The alloys were found to have satisfactory weldability without using fillers or fluxes. The shear strength of the welded joints produced is maintained at a similar level to that of the steel.

4.1.3 (6) Laser welding of kovar to copper

Beranov, M. S. et al. *Weld. Prod.*, 25(5): 17-19, 1978.

It is shown that the strength of the laser welded joint between Kovar and copper corresponds to the strength of annealed copper and the ductility of the joint is maintained. The strength of the weld metal is equal to 92% of that of annealed copper.

4.1.3 (7) Welding of steels and titanium alloys in powerful CO_2 laser

Lebedev, V. K. et al. *Avtom. Svarka*, (4): 30-34, 1979. (In Russian)

Investigates the process of welding of low alloyed steel of some grades and titanium alloys from 3 to 20mm. thick in continuous action laser installations by means of 25kW beams. Data are presented on strength tests and metallographic analysis of welded joints. Laser welding of the steels and Ti alloys up to 20mm. thick at high velocity is considered.

4.1.3 (8) Seam formation features during continuous laser welding

Avramchenko, P. F., Shovkoplyas, V. M. and Velichko, O. A. *Avtom. Svarka*, (4): 61-63, 1980. (In Russian)

To study the power and geometry parameters of laser beam and for direct observation of melting, the effect of laser beam was studied on 70mm. thick organic glass at different depths of focused laser beam. The channels obtained by focusing the beam on the upper and lower surfaces were separated with channels of different depth. The negative angle of front wall profile was also formed during fusion and welding of steel and Ti alloy specimens.

4.1.3 (9) Basic principles and possibilities of application of laser technology in welding technology-II

Dorn, L. and Ohlschlager, E. *Blech Rohre Profile*, 28(4): 152-155, 1981. (In German)

Discusses the effects of the laser beam impinging on the metallic workpiece, particularly reflection and absorption and melting behaviour of laser beams. The properties of laser welded joints and applications of laser welding, such as spot welding and seam welding, are reviewed in detail with numerous examples. CO_2 lasers seem to be the choice in the future for seam welding, although up to 500W, the YAG laser is competitive.

4.1.3 (10) Laser weld attachment of CuNi alloy to ship steel

Banas, C. M. (International Copper Research Association Inc, USA). INCRA Project No. 291, 31pp; 1981. International Copper Research Association, New York, NY, USA.

Laser welding has been suggested as a cost-effective means for attaching CuNi alloy sheet to 1010 ship steel. Lasers are capable of rapid fusion welding with minimum energy input. Narrow fusion zones are obtained with little metallurgical influence on adjacent material. Laser can also be automated and adapted to shipyard requirements, and effectively utilised for seam welding of the clad plates and for loose attachment of the plates to a steel hull. Additional tests were conducted with the specific objectives of: developing appropriate processing parameters for laser weld attachment of 90Cu-10Ni (CDA alloy 706) CuNi to ship steel, obtaining an evaluation of the bonds and generating a set of samples for more extensive mechanical, metallographic and corrosion tests. (7 refs.)

4.1.3 (11) Laser welding of noble metal thermocouple fine wires

Pothoven, F. and Waters, R. In, *ICALEO '82, Proc. Materials Processing Symp.*, Vol. 31, 20-23 September 1982, Boston, MA, pp. 116-119. Laser Institute of America, Toledo, OH, USA, 1982.

The development of a specialised laser system for welding fast response small bead thermocouples fabricated from Nichrome V, Chromel and Pt-10Rh alloys is described. Fast response thermocouple performance is predicated by a small mass thermocouple junction. Until now thermocouple bead diameters have been of the order of three times the wire diameter and have subsequently suffered a slow response time. With the system described, the bead diameter is kept to a nearly 1:1 relationship with the wire.

4.1.3 (12) Laser welding of norsial metallic sandwich structures

Haroutel, J. (Société Nationale Industrie Aerospatiale, France). *Soudage Tech. Connexes.*, 36(1-2): 25-31, 1982. (In French) Also, full report available as rept. No. SNIAS-821-422-144, 22pp; 1981.

Laser welding ensures relatively inexpensive economical fabrication of corrugated metal sandwich structures. The paper describes economic laser welding operations in the manufacture of such structures for air inlets of aerospace vehicles. Mechanical properties of the laser welds are discussed.

4.1.3 (13) Welding metal sandwich structures by laser

Haroutel, J. and Lagoutte, G. *Weld. and Met. Fabr.*, pp. 436-439, 1983.

Norsial is a metal panel with a corrugated core welded between two outer sheets or skins, offering multiple combinations. The laser, associated with numerical control, produces high-reliability rigid Norsial panels. Mechanical characteristics are maintained, and the level of automatic operation considerably reduces the cost of producing Norsial structures.

4.1.3 (14) Joining of iron base P/M parts

Asaka, K. and Hayasaka, T. *J. Mater. Sci. Soc. Jpn.* 21(3): 160-166, 1984. (In Japanese)

Applications of laser welding to Fe base sinters and diffusion bonding for joining between powder and melt-produced materials are described. (11 refs.)

4.1.3 (15) Lasers advance the precision welding art

Elza, D. (Coherent General, USA). *Tool Prod.*, 50(6): 42-44, 1984.

Compared to conventional processes, laser welding offers many productive and quality-improving advantages. No welding rods, fluxes, or protective materials are needed, for example. Dissimilar materials can be joined or sealed as well. Warping, internal stresses, and cracks are minimised because the beam generates heat only in the weld area. The beam can be directed with mirrors so it impinges on surfaces difficult to reach conventionally. Also, a laser can be made fully automatic by computerised numerical control.

4.1.3 (16) Prospects brighter for laser welding

Vanderwert, T. L. (Data Card Corp, USA). *Mach. Des.*, 56(10): 95-100, 1984.

Based on computer numerical control (CNC), laser welders control both the material and laser along up to nine axes at high speeds. Laser welding offers the ability to weld a range of similar and dissimilar materials. Also, the heat affected zone (HAZ) is small. Finally, because laser welding can be automated, it produces consistently good welds.

4.1.3 (17) Users' manual for the laser welding code WELD2D

Russo, A. J. (Sandia National Laboratories, USA). Rept. No. SAND-84-0397. 37pp; April 1984. Sandia National Laboratories, Albuquerque NM, USA.

The two-dimensional laser welding code, WELD2D, was developed to model the conduction mode welding of common metals. For butt welded configurations two dissimilar materials may be used. Either Gaussian or uniform laser beam power distributions may be selected and insulated or conducting ends can be treated. Specification of the laser wavelength, energy per pulse, pulse duration and repetition rate is required as input and the temperature field and molten pool shape are calculated as functions of time. Aluminium, nickel, steel, molybdenum, copper and silicon are included in the code; however, these may be modified or expanded easily with simple changes to data records.

4.2 LASER CUTTING

4.2.1 General

4.2.1 (1) International electron beam processing seminar

Silva, R. M. (Ed.) *Proceedings of 4th Int. Electron Beam Processing Seminar*, April, 1976, Long Island, NY, USA. Universal Technology Corp, Dayton, OH, USA, 1976.

Includes 19 papers dealing with some current industrial applications of electron beams including laser cutting.

4.2.1 (2) Comparative surface integrity study of laser cutting with other conventional cutting techniques

Roy, S. (Imperial College, UK). *Sheet Metal Ind.*, 54(10): 994 and 1003, 1977.

Compares a wide range of cutting methods, including the use of the laser, and their effect on the suitability of blanks cut by the various methods, for subsequent utilisation.

4.2.1 (3) New sheet metal cutting processes using CO_2 lasers

Buness, G. and Weber, S. (Zis, Halle, E. Germany). *Zis Mitt*, 19(1): 120-129, 1977 (In German)

The possibility of modifying the kerf geometry by varying laser beam and technological parameters is demonstrated.

4.2.1 (4) Arc augmented laser cutting

Clarke, J. and Steen, W. M. (Imperial College UK). In, *Proc. Laser '79, Opto-Electron Conf.*, 2-6 July 1979, Munich, W. Germany, pp. 247-253. IPC Science and Technology Press, Guildford, UK, 1979. Also, *Conf. on New Frontiers in Tool Materials, Cutting Techniques, Metal Forming*, 15-16 March 1979, London, Session V, Paper 1. Sponsored by Engineers Digest.

The two energy sources cooperate to give enhanced thermal effects and the mechanism of this coupling is discussed by a comparison of experimental results with theoretical calculations from a mathematical model. It is shown that this cutting technique allows the possibility of a laser of given power achieving the performance expected of a laser having the power equivalent to both arc and laser combined. At present, a 60% increase in the cutting speed of a 2kW CW CO_2 laser has been achieved by the addition of 2kW of developed arc power. (15 refs.)

4.2.1 (5) The laser as an industrial cutting tool

Hoffmann, M. *Met. Constr.*, 11(1): 33-34, 1979.

The characteristics and advantages of laser cutting and selection criteria favouring the introduction of laser cutting are reviewed. Laser fusion cutting, laser oxygen cutting, and the popular use of CO_2 lasers are considered. The main advantages of the process include its efficiency, low operating costs, robust design of equipment, facility for automation, and precision.

4.2.1 (6) The CO_2 laser in industrial applications

Scheuermann, W. *Ind. Anz.*, 102(73): 49-53, 1980. (In German)

The economic feasibility of a laser cutting plant is demonstrated by work time reduction, non-contact and quiet cutting operation with a multiplicity of materials, narrow kerf, high cutting speed, minor heat influence on the border zone, no material distortion, reduction in marking and, in most cases, in reworking manufacture of complex form cutting.

4.2.1 (7) Chipless forming – Today and tomorrow

Proc. Chipless Forming – Today and Tomorrow Conf., 30-31 March 1982, Birmingham, UK. Fuel and Metallurgical Journals Ltd, Redhill, UK, 1982.

Proceedings contain 13 papers, including coverage of laser cutting.

4.2.1 (8) Cutting materials with the aid of CO_2 lasers

Kosyrev, F. K. et al. *Avtom. Svarka*, (3): 68-70, 1982. (In Russian)

Experiments with steel Kh15N5D2T and 30KhSND, Al alloy, tungsten, Mo, rubber, plastic and glass fibre confirmed the possibility of high-speed, high-quality cutting of materials using CO_2 lasers with power of 0.8 and 4.9kW. Briefly describes a machine for laser cutting with numerical program control, ensuring cutting of parts from thin sheet steel with a precision of 0.3mm.

4.2.1 (9) Laser cutting – a progressive technology in sheet metal working

Benzinger, M. and Weick, J. *Ind. Anz.* 104(78): 36-41, 1982. (In German)

Presents the background technology for laser cutting, covering types of CO_2 lasers and cutting characteristics, influence of size on metal cutting, qualitative characteristics of laser cuts and production technology of laser cutting. Laser cutting combined with punching is recommended for its effectiveness.

4.2.1 (10) Qualitative and economic aspects of materials processing by means of CO_2 laser systems

Herbrich, H. In, *Colloquium on Thermal Cutting and Flame Processes*, 7 September 1982, Ljubljana, Yugoslavia, 9pp. International Institute of Welding, London, UK, 1982.

Describes a fast axial flow CO_2 gas laser which has improved beam characteristics and capacitive distance control of the processing optics. Economic aspects of laser cutting are considered by comparing the costs of gas cutting, plasma-arc cutting and 500W and 1000W laser cutting. Laser cutting is particularly favoured for two-or three-dimensional shape cutting.

4.2.1 (11) Industrial applications of laser cutting

Schekulin, K. *Deformacion Met.*, (87): 49-55, 1983. (In Spanish)

Laser cutting is briefly treated and a commercial numerical controlled laser cutting device described. Various examples of pieces cut by a laser are then given. Laser cutting is also ideal for short series runs and during development work, as well as for large series production runs. Laser cutting can be usefully employed when blanking is difficult, and for plastics.

4.2.1 (12) Reducing the concentration of pollutant substances during thermal cutting

Becker, W. *Schweissen und Schneiden*, 35(7): 303-308, 1983. (In German)

The pollutant substances which may be produced during thermal cutting of Al and steel and the statutory regulations relating to conditions at the workplace are discussed. There are three possible methods of reducing the concentration of pollutant substances at the workplace to permissible levels: water spraying for the oxy-fuel gas cutting of workpieces up to 120mm. thick, air extraction for plasma cutting, oxy-fuel gas cutting of thicker workpieces and laser cutting and a water table for wet plasma cutting.

4.2.1 (13) Laser materials cutting and related phenomena

Berloffa, E. H. and Witzmann, J. (Voest-Alpine AG, Austria). In, *Proc. SPIE Industrial Applications of High Power Lasers*, Vol. 455, 26-27 September 1983, Linz, Austria, pp. 96-101. SPIE International Society of Optical Engineers, USA; Austrian Physics Society, et al, 1984.

4.2.1 (14) Advanced cutting techniques: laser and fissuration cutting – II Laser beam cutting

Migliorati, B.(Fiat TTG, Italy). In, *Decommissioning of Nuclear Power Plants, Proc. of European Conf. EUR 9474*, Luxemburg, 1984, (Eds. K. H. Schaller and B. Huber), pp. 244-251. Graham and Trotman, London, UK, 1984.

Experimental tests have been performed using a CO_2 laser with output power 1-15kW to evaluate the effect of varying the following parameters: material (carbon steel Fe 42 C, stainless steel AISI 304, concrete), laser power, beam characteristics, workpiece velocity, gas type and distribution on the laser interaction zone. In the case of concrete, drilling depths of 80mm. were obtained in a few seconds using a 10kW laser beam, and pieces of 160mm. were cut at 0.01m per minute. Results with carbon steel indicated maximum thicknesses of 110mm, cut at 0.01m per minute with 10kW, depths about 20% lower were obtained with the AISI 304 stainless steel. A parallel investigation was aimed at characterising particulate emission during the laser cutting process.

4.2.1 (15) Laser or plasma cutting

Fischer, R. *Schweiz. Maschinenmarkt*, 84(32): 16-19, 1984. (In German)

Both cutting methods depend on producing high temperature, whereby the steel to be removed is melted or vapourised. A high-velocity gas stream drives the melt out of the cut; if inert gas is used, the material is removed thermally, whereas use of an oxygen blast results in exothermic reaction. The advantages relative to shearing are higher cutting speeds, better quality of cut, no cutting tool required and no force acting on the sheet. For laser cutting, power output must be sufficient and the beam must be in the ground mode. These requirements are met by a CO_2 laser of 250-1250W, capacity focused to 0.1mm. diameter.

4.2.1 (16) On the maximum limit value in the cutting speed with a plane polarised laser beam

Lepore, M. (Bari University, Italy). *Appl. Phy. Commun.*, 4(2-3): 121-134, 1984.

4.2.1 (17) A quantitative theory for the role of oxygen in the laser cutting process

Lepore, M. et al. (Centro Laser, Italy). In, *ICALEO '83, Proc. Materials Processing Symp.*, Vol. 38, 14-17 November 1983, Los Angeles, CA, (Ed. E. A. Metzbower), pp. 160-165. Laser Institute of America, Toledo, OH, USA, 1984.

A quantitative theory giving a unified view of certain laser applications was developed. This theory distinguishes two regimes in the oxygen assisted laser cutting process (OALCP): laser alone and laser + oxygen reaction. Further experimental evidence which supports the theory, was obtained by means of polarised laser radiation. (15 refs.)

4.2.1 (18) Guidelines for laser cutting

Darchuk, J. M. and Migliore, L. R. (Spectra-Physics Industrial Laser Division, USA). *Lasers and Appl.*, 4(9): 91-97, 1985.

Reviews the basics of laser cutting.

4.2.1 (19) How and why a laser?

Laos, O. V. (Laser Scientific Services Ltd, UK). *Electronics Power*, 31(5): 381-383, 1985.

Discusses the factors which should be considered before purchasing laser cutting equipment.

4.2.1 (20) Introduction to laser cutting

Watson, M. N. (Machine Tool Industry Research Association, UK). Rept. No. MTIRA-85/01/XAD, 69pp., April 1985. (Prepared in cooperation with Welding Institute, Cambridge, UK).

After a general introduction to lasers for material processing, looks at applications of lasers to cutting and the machine tools. Then reviews the hardware commercially available, including solid state (YAG) and CO_2 lasers. Finally, a section is included on safety precautions to be used when working with lasers.

4.2.1 (21) Laser cutting

Beyer, E. et al. (Fraunhofer-Institut fur Lasertech., Germany). *Laser & Optoelektron.*, 17(3): 282-290, 1985. (In German)

Briefly compares CO_2-Nd:YAG and excimer laser material interactions; the cutting process by CO_2 and Nd:YAG lasers is discussed in more detail: cutting by sublimation, melting and oxygen reaction. Focusability, absorption and reflection properties and melting dynamics are dealt with, and examples of CO_2 laser cutting are presented. (15 refs.)

4.2.1 (22) Relationships governing evaporation of particles in vacuum under the effect of laser radiation

Popova, L. V. and Sutugin, A. G. *Phys. & Chem. Mater. Treat.*, 19(3): 178-180, 1985.

Theoretical investigations were carried out into the evaporation of particles leaving the metal in vacuum laser cutting. Calculations were conducted to determine the evaporation time and the temperature of spherical aerosol particles of graphite and aluminium in relation to their optical parameter, radiation flux and also radiation wavelength.

4.2.1 (23) Why a laser cutting system: a comparison with more conventional cutting systems

Graham, R. E. (General Fabrication Dept., US Amada, USA). In, *Proc. SPIE Applications of High Power Lasers*, Vol. 527, 22-23 January 1985, Los Angeles, CA, pp. 6-10. SPIE International Society Optical Engineers, USA, 1985.

Marketing a laser cutting system usually requires a comparison with plasma-arc cutting, wire electrical discharge machining, and oxygen-acetylene flame cutting. Each system has its own advantages and disadvantages. However, because of it versatility and adaptability, laser is becoming the more sought-after system, as costs come down.

4.2.1 (24) Reactive gas-assisted cutting with pulsed Nd: YAG lasers

Van Dijk, M. H. H. and Brouwer, E. (Twente University, The Netherlands). In, *Laser Welding, Machining and Materials Processing – Proc. Int. Conf. on Applications of Laser and Electro-Optics, ICALEO '85*, 11-14 November 1985, San Francisco, CA, (Ed. C. Albright), pp. 153-162. Laser Institute of America, Toledo, OH, USA/IFS (Publications) Ltd, Bedford, UK, 1986.

The maximum speed in gas-assisted laser cutting depends on several parameters. The influence of the pressure on the surface of the workpiece and the position of the focal point has been investigated with the use of a pulsed Nd:YAG laser. A procedure to find the optimum setting of pulse power, pulse length and pulse frequency is given.

Experimental results are compared with the Line Source Model (LSM).

4.2.2 Laser cutting of metals

4.2.2 (1) Laser cutting replaces blanking

Schaffer, G. *Am. Machinist*, 122(12): 100-103, 1978.

Discusses the use and advantages of auto-mated NC laser cutting of stainless steel sheet components. This process was developed for the manufacture of stainless steel plate-and-fin heat exchanger cores because of improved economics and flexibility.

4.2.2 (2) Production laser cutting

Huber, J. and Marx, W. (Grammen Aerospace Corp, USA). In, *Proc. Applications of Lasers in Materials Processing Conf 18-20 April 1979, Washington DC, USA, pp. 273-290. American Society for Metals, OH, USA, 1979.*

Initial production applications on steel and Ti alloy parts were accomplished with a 250W, continuous wave, CO_2 laser. Laser cutting has reduced rough trimming costs by 60 to 80%. Production laser cutting parameters, such as power, feed rate, assist gas, focal spot size, operating costs and safety requirements for the existing equipment have been established. Control requirements and equipment modifications are discussed for laser cutting both flat and contoured parts.

4.2.2 (3) Technology and equipment for gas laser cutting of thin steel sheet

Tikhomirov, A. V. et al. *Weld. Prod.*, 26(11): 11-13 1979.

A CO_2 laser with power to 1kW was used for cutting high-strength steel VNS2 and Kh18N10T stainless steels. The device was used by an automobile building plant for making precision parts, non-ferrous metals cutting, glass-plastics annd composite materials cutting.

4.2.2 (4) Laser cutting is a job shop specialty

Jefferson, T. B. *Weld. Des. Fabr.*, 53(3): 109-111, 1980.

Describes the use of a CO_2 laser cutter to make parts from stainless steels, Al, Ti, exotic metals and alloys. Since the medium power (525W) laser cuts rapidly at a low cost/in., it is especially suitable for short-run stamping, piercing and drilling. The laser has two cutting modes, continuous wave and pulse, which doubles power output. Accuracy is ±0.003in. in 48in. The laser is fitted with a proximity sensor to ensure proper focal length and clean, sharp cuts.

4.2.2 (5) Laser cutting of flat cross-sections

Vanschen, W. *Blech Rohre Profile*, 27(9): 545-549, 1980. (In German)

Presents an introduction to recent developments of carbon dioxide laser beam cutting equipment. Three-dimensional cutting processes are possible. The emphasis of the applications is on thin sheet metal cutting.

4.2.2 (6) Laser cutting solves complicated fabrication problem

Bohme, O. et al. *ZIS Mitt.*, 22(3): 304-307, 1980. (In German)

Discusses cutting slots in the side and columnar filters used in the chemical processing industry. The filter structures resembled a double-helix extruder. The slots had to be 0.6mm wide. A power laser with flexible cutting head was used with power output of approximately 160W. A cutting speed of 800mm/min. was reached. The width of the slot cut by the laser was 0.28mm.

4.2.2 (7) Lasers cut a path in sheet metalworking

Astrop, A. *Mach. Prod. Engng.*, 137(3535): 17-19, 1980.

Desitech Ltd of Wellingborough, UK, has produced a high-speed CNC profiling and point-to-point machine with laser cutting tool. The laser system is of CO_2 type with power outputs from the 100 to 425W range, and on continuous rating typically provides 375W. A feature of the system, however, is the ability to switch to an 'enhanced pulse', where a peak power of 2000W is available over pulse lengths up to 100μs and at pulse frequencies up to 2.5kHz.

4.2.2 (8) The importance of gas flow parameters in laser cutting

Kamalu, J. N. and Steen, W. M. In, *Metallurgical Society AIME*, Warrendale, PA, USA, 1981, Paper No. A81-38, 20pp. Also in *Lasers in Metallurgy; Metallurgical Society/AIME*, pp. 263-278, 1981.

Nozzle design and cutting gas flow characteristics play an important role in the laser cutting process and need to be optimised for a given cutting operation. The optimisation can be effected by relating such factors as nozzle diameter and cutting gas exit pressure/velocity distribution to actual cutting performance results. Cutting performance is assessed in terms of cutting speed and cut quality. It is shown that for a given power input, optimum cutting per-

formance is dependent on nozzle design and cutting gas pressure. Laser cutting of 2mm. gauge mild steel specimens is discussed. (10 refs.)

4.2.2 (9) Laser metal cutting

Gregson, V. G. *Mach. Tool Blue Book*, 76(9): 72-75, 1981.

Typical cutting rates are 140in./min for 0.040in. thick stainless steel, 100in./min for 0.250in. Ti and 150in./min for 0.250in. acrylic. Laser cutting provides small heat affected zones, flexibility, minimum material waste, reduced secondary operations and high precision for small holes, narrow slots and closely spaced patterns. The laser process is described and factors are given for choosing a laser.

4.2.2 (10) Process alternatives: nibbling, laser cutting and plasma arc cutting

Lunn, D. J. *Tool. Prod.*, 46(12): 66-69, 1981.

Laser cutting in sheet metal work is more accurate than plasma arc, but although leaving square-cut edges on both sides of the 0.005-0.008in. diameter beam, it cannot approach the high cutting speeds of plasma arc. Its main advantage is for intricate cutting, such as for precision electronic panels and surgical instruments.

4.2.2 (11) Cutting and edge preparation for welding

Weld. Rev., 1(1): 26, 29-30, 33-34, 1982.

Briefly describes the three most common processes used: oxy-fuel gas cutting, plasma arc cutting and laser cutting. In laser cutting, metal melted by the absorption of radiant heat from a finely focused monochromatic light beam is blown away by an ejected gas.

4.2.2 (12) Cutting steel sheets with a CO_2 laser

Tamaschke, W. *Werkstattstechnik*, 72(6): 323-326, 1982. (In German)

Experiences with cuttable materials including St12 and X8Cr17 and a machine for combined stamping and laser cutting operations and a laser cutting machine are described. Safety considerations are discussed.

4.2.2 (13) Gas laser cutting of materials

Kudrayavtsev, E. P. and Tikhomirov, A. V. In, *Colloquium on Thermal Cutting and Flame Pro-* cesses, (included in the Annual Assembly of the International Institute for Welding), 7 September 1982, Ljubljana, Yugoslavia, 9pp. International Institute for Welding, London, UK, 1982.

Considers estimated and experimental conditions of gas laser cutting of materials when employing CO_2 lasers up to 5kW capacity; gives analysis of the designs of machines intended for gas laser cutting of metal sheets, and present estimation and experimental results on the operating accuracy of a numerical controlled gantry machine.

4.2.2 (14) The laser – an alternative to the punch press

Field, R. *Hydraul. Pneum, Mech. Power*, 28(328): 134-135, 1982.

Physical characteristics of the laser beam and its operation are discussed to assess the benefits to be gained by laser cutting. The beam can be transmitted up to 20ft without loss of power or definition. Difficult materials can be cut with a very narrow cut width, and a burr-free cut is produced. Comparable cutting speeds are tabulated that can be achieved with 1/2kW and 1kW lasers and various thicknesses of standard mild steel. A range of materials that can be cut is given, with comparative cutting rates; examples are Ti, heat treated steels, plastics, and Al.

4.2.2 (15) Laser applications in sheet metal working

Toenshoff, H. K. and Balbach, J. (Technical University, Hannover, W. Germany) *Baender Bleche Rohre*, 23(9): 246-249, 1982. (In German)

Describes the portal type of laser cutting machine with which the beam device is moved over the workpiece, and the model with a stationary laser unit and a coordinate table. Materials suited to this process are steels – mainly unalloyed – and titanium. The laser beam enables practically any shape to be produced to high standards.

4.2.2 (16) Laser cutting of sheet metal ranging in thickness from 3 to 30mm

Rothe, R. et al. In, *Schweissen und Schneiden '82*, 29 September-1 October 1982, Berlin, Germany, pp. 40-48. (In German), Deutscher Verlag fur Schweisstechnik, Dusseldorf, W. Germany, 1982.

Studies were made on laser sublimation cutting, laser fusion cutting and laser flame cutting of Al and steel, mostly with 4kW lasers. Up to 30mm. sheets can be cut using this process.

4.2.2 (17) Laser cutting of steel with the revolutionary turbolase T1000 CO₂ laser

Eckersley, J. S. In, *ICALEO '82, Proc. Materials Processing Symp.*, Vol. 31, 20-23 September 1982, Boston, MA, pp. 131-134. Laser Institute of America, Toledo, OH, USA, 1982.

The Turbolase T1000 is perhaps the first laser specifically designed for metal cutting, employing Photon's unique Turbotube that delivers 1000W of laser energy to produce smooth, straight cuts with no more than four optical elements in the entire cutting system. Its design and use in the cutting of Al, carbon, stainless and galvanised steel are discussed.

4.2.2 (18) Lasers speed sheet metal progress

Metalwork. Prod., 126(8): 81-82, 1982.

During 1980 Mason and King Ltd, (Leicester, UK), installed a Trumpf Laserpress 180LK laser cutting machine, and have now added a 180LW model with automatic tools changing. These are claimed to give a better standard of finish on profiled sheet metal straight off the machine and profiling at speeds up to 10m/min. Programming is carried out tool-by-tool, first punching, then providing start-up holes, and then laser profiling.

4.2.2 (19) Quality improvement in CO₂ laser cutting of metals

Hoshinouchi, S., Kobayashi, M. and Shimada, W. In, *Colloquium on Thermal Cutting and Flame Processes*, 7 September 1982, Ljublijana, Yugoslavia, 6pp. International Institute of Welding, London, UK, 1982.

Investigates the effect of various process variables on quality with both continuous wave and pulse wave CO₂ lasers. It is explained that the control of heat input is the most important if high quality cutting is to be achieved.

4.2.2 (20) Seminar on efficient metal forming and machining, 1982

Proc. Seminar on Efficient Metal Forming and Machining, 16 November 1982, Pretoria, S. Africa, CSIR (S 290), Pretoria, S. Africa, 1982.

Contains 18 papers whose main topics include laser cutting. Also covered are topics on industrial economics and the impact of new technologies on the metal forming industry.

4.2.2 (21) Stainless steel plate cutting: thermal and mechanical cutting — III Laser cutting — some observations

Balbi, M. and Silva, G. (Milan Polytechnic, Italy). *Avesta Stainless Bulletin*, 6(4): 3-11, 1982.

The characteristics of laser cutting are linked with energy concentration, which makes it possible to work at high cutting rates with a very narrow cut, limited specific heat supply and, consequently, a small HAZ with no heat distortion. It is especially suited for complex, contoured cuts on small areas and sharp edged bevels. The part is not mechanically stressed during the cutting operation and the plates cannot warp.

4.2.2 (22) Temperature controlled laser cutting

Lenz, E., Shachrai, A. and Tal, Y. In, *ICALEO '82, Materials Processing Symp.*, Vol. 31, 20-23 September 1982, Boston, MA, pp. 149-161. Laser Institute of America, Toledo, OH, USA, 1982.

A project carried out at the Israel Institute of Technology, where a complete on-line adaptive control system was built to optimise the laser cutting process, is described. The adaptive control constraint (ACC) is based on the experimental results, indicating a significant correlation between the kerf quality and the workpiece temperature at a close vicinity to the cut front. Temperature measurement method, cut quality and surface temperature correlation, the adaptive control system and experimental results are discussed. The experiments were carried out on AISI 304 stainless steel 3mm. thick plates using a CO₂ laser.

4.2.2 (23) Use of the laser in sheet metal working

Balbach, J. and Tonshoff, H. K. *Bander Bleche Rohre*, 23(9): 246-249, 1982. (In German)

Briefly describes the portal type of laser cutting machine with which the beam device is moved over the workpiece and the model with stationary laser unit and coordinate table. Metals suited to this process are unalloyed steels and Ti. The laser beam enables practically any shape to be produced to high standards of quality.

4.2.2 (24) Advanced development and production of punching and cutting systems

Sheet Met. Ind., 60(3): 152-154, 156, 159, 1983.

The contribution made by the Trumpf group in the development and promotion of high tech-

nology punching and nibbling systems is described. Machines described and illustrated include a laser press for laser cutting.

4.2.2 (25) An investigation of the cutting process with the aid of a plane polarised CO_2 laser beam

Opt. Lasers Engng., 4(4): 241-251, 1983.

A plane polarised power beam was used to investigate the reactive gas-assisted laser cutting process. Gives further experimental support for the theory of the existence of two different regimes in the oxygen-assisted cutting of steels and explains the dependence of the process on FeO_2 exothermic reactions. (8 refs.)

4.2.2 (26) Laser cutting

Steen, W. M. and Kamalu, J. N. (Imperial College, UK). In, *Material Processing Theory and Practice Vol. 3, Laser Material Processing*, pp. 15-111. North – Holland Publishing Co, Amsterdam, 1983.

Discusses the laser cutting process by first describing the physical mechanism whereby a laser is able to cut, then illustrating how it is currently used in industry and some of the current research developments which are being considered. This is followed by a section on the practical performance of CW lasers in cutting, with tables and graphs of the relationships between the principal laser operating parameters and the cut speed and quality. The operating parameters considered are: laser power, beam diameter, mode structure, F number of focusing optics, nozzle design, gas flow and composition, and arc augmentation. Pulsed lasers are treated separately together with the specialist cutting processes such as hole drilling, etching, scribing and resistance trimming.

4.2.2 (27) Making prototype stampings by laser

Vanderwert, T. L. (Data Card Corp, USA). *Mach. Des.*, 55(23): 107-111, 1983.

The equipment can produce prototype stampings at substantially less cost and without the delay of conventional blanking methods. Furthermore, design changes can be accommodated simply by changing the 'tooling' which consists of a NC program. Laser machining has the capability to produce a limited number of stampings while permanent dies are being fabricated.

4.2.2 (28) Industrial practice in laser cutting of three-dimensional formed sheet parts

Herbrich, H. In, *Proc. Sheet Metal Working '84 Conf.*, 7-8 November 1984, Essen, W. Germany, pp. 163-174. VDI Verlag GmbH, Dusseldorf, W. Germany, 1984.

CO_2 gas lasers with 200-1000W power are usually used for cutting, showing the least HAZ of all thermal cutting processes. Cutting rates of 2m/min. are attainable in 4mm thick St37 sheet with a 1000W CO_2 laser, with cut widths in the range 0.1-0.8mm. Crucial to quality of cut is maintenance of the 1mm. distance between the laser cutting nozzle and the work surface. This is maintained within a range of ±0.2mm. by use of pickups utilising capacitative measurement between electrodes surrounding the laser nozzle and the work surface. The use of five-axis numerically controlled workpiece and laser beam manipulators facilitates the cutting of workpieces of complex geometric profile.

4.2.2 (29) Laser beam cutting of thick steel

Sepold G. and Rothe, R. (BIAS Bremer Institut für angewandte Strahltechnik, Germany). In, *ICALEO '83, Proc. Materials Processing Symp.*, Vol. 38, 14-17 November 1983, Los Angeles, CA, (Ed. E. A. Metzbower), pp. 156-159. Laser Institute of America, Toledo, OH, USA, 1984.

The exothermic reaction heat provides a high fraction of the total process power. Therefore it is possible to cut steel plates more than 30mm. thick with a 5kW laser. Experiments had been carried out cutting mild steel and austenitic Cr-Ni-steel. The cutting set-up and some results are shown. The results are discussed and compared with the results of conventional oxy-flame cutting.

4.2.2 (30) Laser profiling in the metalworking industry

Wilhelm, H. In, *Proc. Developments and Innovations for Improved Welding Production – 1st Int. Conf.*, 13-15 September 1983, Birmingham, UK, pp. 16.1-16.6. Welding Institute, Cambridge, UK, 1984.

The CO_2 laser cutting process is introduced as one of the main thermal cutting processes. Various methods of designing laser profile cutting machines are explained and their individual merits and disadvantages discussed. The benefits of using the laser for metal profiling are explained and comparisons are drawn with other production methods. (2 refs.)

4.2.2 (31) Manufacturing of laminated deep drawing dies by laser beam cutting

Kunieda, M. and Nakagawa, T. In, *Advanced Technology of Plasticity*, Vol. 1, pp. 520-525. Japan Society for Technology of Plasticity, Tokyo, Japan, 1984.

If a number of sheets with gradually changing profile dimensions cut on them are laminated, a three-dimensional arbitrary curved surface will be generated, which will be stepped or discrete. In a similar manner, if a known form of die or punch surface is divided by equally spaced planes and the known profiles on these slices are cut on sheets of same thickness and then assembled, the form desired can be achieved but with steps. By grinding, these steps can be removed to give the surface of the die. This method was tried to manufacture dies and punches which revealed that both the delivery time from the date of order and cost of manufacturing can be reduced drastically if the sheet profiles are cut by laser beam. The results showed great scope of realising FMS in the field of plastic working and mold forming processes.

4.2.2 (32) Metalworking's shopfloor ally

Atkey, M. *Mach. Prod. Engng.*, 142(3656): 18-19, 21-22, 24, 1984.

In addition to the inherent advantages of cutting with light, the laser now has the reliability, performance and power for effective shopfloor use. Two types of laser predominate: carbon dioxide and yttrium aluminium garnet (YAG). The small diameter beam and high-pulsed powers make YAG an ideal tool for drilling/trepanning type work in composite alloys in the aero-engine sector. CO_2 lasers, with their speed and cutting power, have found applications in the motor vehicle industry.

4.2.2 (33) Use of laser cutting

Wellendorf, K. *Schweisstechnik*, 34(3): 111-113, 1984. (In German)

Describes a CNC laser cutting system and the associated operator workplace. Specific laser cutting technologies and the satisfactory cutting results obtained for carbon and alloy steels are proof of the system's high operational efficiency.

4.2.2 (34) Application of lasers in sheet metal production

Hahn, W. *Ind. Anz.*, 107(38): 28-31, 1985. (In German) Also, *Werkstadt. Betrieb.*, 116(6): 381-383, 1983. (In German)

The present generation of laser cutters of the TRUMATIC series is available with cutting pow-

ers between 600 and 1250W with cutting speeds of up to 15m/min for thin sheets. The maximum sheet thickness is 10mm. The accuracy of laser cutting is compared with conventional methods, such as punching and nibbling, and applications in complex geometrical parts cutting is demonstrated.

4.2.2 (35) CO_2 laser cutting process of thick aluminium

Dell'Erba, M., Daurelio, G. and Ferrara, M. (Centro Laser, Italy). *Appl. Phys. Commun.*, 5(1-2): 23-35, 1985.

Reports the experimental results of laser cutting of aluminium, obtained with a 2kW CW CO_2 laser. The data analysis, by means of the Line Source Model, has singled out a behaviour different from the one previously found for steels and other metals. In this way, two experimental equations for the laser cutting process of metals are obtained.

4.2.2 (36) Cut edge quality improvement by laser pulsing

Powell, J. et al. (Loughborough University of Technology, UK). In, *Proc. 2nd Int. Conf. on Lasers in Manufacturing*, 26-28 March 1985, Birmingham, UK, (Ed. M. F. Kimmitt), pp. 37-46. IFS (Publications) Ltd, Bedford, UK, 1985.

Analysis of the CO_2 laser oxygen jet cutting of mild steel has led to the development of a new cutting technique utilising a specific range of laser pulsing frequencies. Profilometry and SEM analysis of the cut edges have shown remarkable improvements in quality; Ra values of the surface roughness have been reduced by a factor of three or more.

4.2.2 (37) Cutting of aluminium plate with medium output CO_2 laser

Murakawa, M. et al. (Nippon Institute of Technology, Japan). *Bulletin Jpn. Soc. Precis. Engrs.*, 19(3): 223-224, 1985.

Describes the experimental results of cutting an aluminium plate by CO_2 laser machine with enhaced pulse feature to find the most appropriate combination of cutting parameters.

4.2.2 (38) Laser beam cutting of metallic and structural components

Manassero, G. et al. Commission of the European Communities, Luxembourg, 1985, 27pp. (In Italian)

Experimental tests have been performed using CO_2 laser with output power 1-15kW to evaluate the effect of varying material (carbon steel Fe42C, stainless steel AISI 304, concrete), laser power, beam characteristics, workpiece velocity, gas type and distribution on the laser interaction zone. In the case of concrete, drilling depths of 80mm. were obtained in a few seconds using a 10kW laser beam, and pieces of 160mm were cut at 0.01m/min. Results with C steel indicated maximum thicknesses of 110mm, cut at 0.01m/min with 10kW, while depths of approx 20% lower were obtained with the AISI 304 stainless steel.

4.2.2 (39) Laser cutting of aluminium thin film with no damage to under-layers

Mitani, M. et al. (Hitachi Ltd, Japan). *Ann. CIRP*, 28(1): 113-115, 1979. Updated version in *IEEE Proc. Solid State Circuits*, SC-20(6): 1259-1264, 1985.

4.2.2 (40) Laser cutting of sheet metal in modern manufacturing

Weick, J. M. and Wollermann-Windgasse, R. (Trumpf GmbH & Co, West Germany). In, *Proc. 2nd Int. Conf. on Lasers in Manufacturing*, 26-28 March 1985, Birmingham UK, (Ed. M. F. Kimmitt), pp. 47-56. IFS (Publications) Ltd, Bedford, UK, 1985.

The state-of-the-art of laser cutting is described. Cutting parameters for 500W, 750W and above 1000W and typical applications for the different power levels are demonstrated. Limits, caused by material thickness, material composition and geometry of the workpiece are discussed. The advantages of pulse mode for higher accuracy are shown. The benefits of the combination punching/laser cutting for the manufacturing process and the gain of flexibility are shown by discussing some typical workpieces.

4.2.2 (41) Laser cutting of titanium materials

Decker, I. et al. *Schweissen und Schneiden*, 37(8): 356-362, 1985. (In German) (Shortened version in English pp. E117-E119).

Discusses the influence of the cutting parameters on the geometric condition of the cut surface and on changes in the microstructure of the heat-affected surface layer during thermal cutting of pure Ti and Ti6Al4V using a laser beam. The differences between laser beam cutting with oxygen and laser beam fusion cutting are clarified. In the process, a connection is established between the available laser power and the cutting efficiency. (17 refs.)

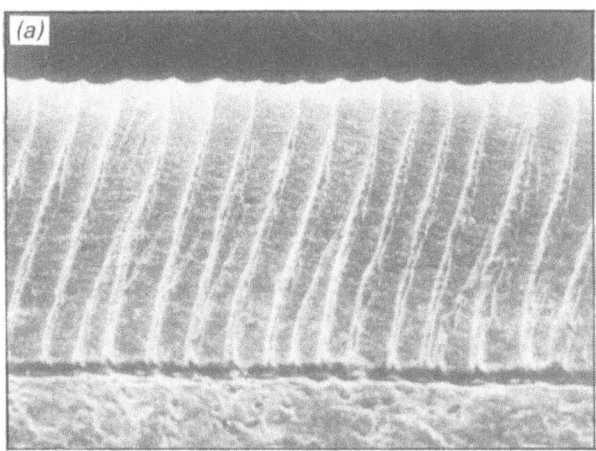

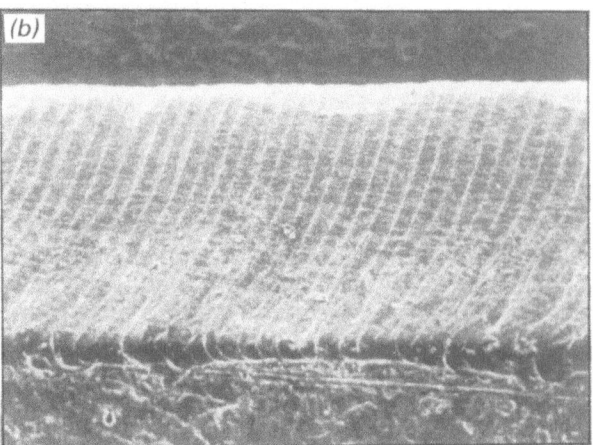

(a) Scanning electron micrograph of typical striated laser cut edge of a mild steel specimen, and (b) SEM of improved quality cut edge as a result of pulsing at a frequency of 500Hz (See 4.2.2 (36))

4.2.2 (42) On cutting and penetration welding processes with high power lasers

Cingolani, A., Daurelio, G. and Esposito, C. *Opt. Lasers Engng.*, 6(1): 1-9, 1985.

Cutting and continuous welding laser processes are examined. Experiments with CO_2 and YAG lasers were carried out on carbon and stainless steels. Two distinct regimes were identified in the gas-jet-assisted cutting process and in both cases predictions of working parameters can be made. Penetration welding results, when represented on a mathematical model, were seen to be similar to those of cutting ones. (8 refs.)

4.2.2 (43) Parametric studies of pulsed laser cutting of thin metal plates

Lee, C. S., Goel, A. and Osada, H. (Amada Engineering, USA). In, *ICALEO '84, Proc. Materials Processing Symp.*, Vol. 44, 12-15 November

Multiple drossjet arrangement (See 4.2.2 (44))

1984, Boston, MA, (Ed. J. Mazumder), pp. 86-93. Laser Institute of America, 1985. (Also *J. Appl. Phys.*, 58(3): 1339-1343, 1985.)

Reports a parametric study of pulsed laser cutting of thin metal plates. The physical processes of laser-material interaction involving material removal are explored through a systematic study of pulsed laser cutting of various metal plates of different physical properties. The role of assist gas in cutting process was also investigated. (11 refs.)

4.2.2 (44) Prevention of dross attachment during laser cutting

Birkett, F. N., Herbert, D. P. and Powell, J. (Loughborough University of Technology, UK). In, *Proc. 2nd Int. Conf. on Lasers in Manufacturing*, 26-28 March 1985, Birmingham, UK, (Ed. M. F. Kimmitt), pp. 63-66. IFS (Publications) Ltd, Bedford, UK, 1985.

When certain materials are cut using medium power (≈ 500W) carbon dioxide lasers, dross is formed which adheres strongly to the underside of the cut material and which requires an additional removal process. The paper describes a process which gives dross-free cut components when used with stainless steel and similar materials.

4.2.2 (45) Punch presses lead in flexibility stakes

Cookson, J. *Metalwork Prod.*, 129(4): 108-120, 1985.

Advantages and limitations of different types of plasma cutting systems are compared, as well as the particular features of CNC-controlled laser cutting.

4.2.2 (46) Sheetmetal, plate, feel laser's impact

Mach. Tool Blue Book, 80(10): 82-86, 1985.

Argues that the laser now has the reliability, performance and power for effective shopfloor use. Describes the types of lasers predominant in metal cutting: carbon dioxide and yttrium aluminium garnet. The essential differences are based on the lasing medium, the method of beam excitation and the nature of wave output.

4.2.2 (47) State-of-the-art in the laser cutting of metals

Teske, K. *Schweisstechnik*, 39(5): 70-75, 1985. (In German)

Attention is given to the forms of cutting, the various types of cutting laser, the cutting heads and cutting nozzles, guidance equipment and other equipment. Cutting rates and performances are compared, and the problems encountered with certain metals and non-metals discussed. The cutting costs of different lasers are analysed, and laser cutting applications suggested.

4.2.2 (48) Optimisation of pulsed laser cutting of mild steels

Powell, J. et al. (Loughborough University of Technology, UK). In, *Proc. 3rd Int. Conf. on Lasers in Manufacturing*, 3-5 June 1986, Paris, France, (Ed. A. Quenzer), pp. 67-76. IFS (Publications) Ltd, Bedford, UK, 1986.

Previous work has shown great improvements in cut edge finish by judicious modulation of laser output during laser-O_2 cutting of thin section mild steel. This work extends the ex-

perimental range to thicker material with increased laser power. Uniformity of cut edge profile, a feature of the work with thin materials, was found to be disrupted by the fluid flow effects of material ejection from the bottom of the cutting zone. In spite of this added complication, cut edge finish was still generally found to be superior when the laser output was modulated during cutting.

4.2.2 (49) Laser cutting with high pressure cutting gases and mixed gases

Nielsen, S. E. (Technical University of Denmark). In, *Proc. 3rd Int. Conf. on Lasers in Manufacturing*, 3-5 June 1986, Paris, France, (Ed. A. Quenzer), pp. 25-44. IFS (Publications) Ltd, Bedford, UK, 1986.

Describes the development of a high pressure gas-assisted laser cutting head based on a 3-lens system. The effective 'cleaning' of the cut kerf, using high gas pressures, results in increasing cutting velocity and/or quality dependent on the material to be cut.

Experimental studies in the use of different gas mixtures as assisting cutting gas in the laser cutting process are outlined. When cutting stainless steel using oxygen as assisting cutting gas the material is burned away forming different types of chromium-oxides (slag). This slag causes problems when laser cut specimens are to be welded, deburred or surface modified. When using special mixed cutting gases, slag-free cut kerfs appear without significant reductions in the cutting velocity.

4.2.3 Laser cutting of non-metals

4.2.3 (1) Applications for industrial lasercutter systems

Hanson, W. E. (Hughes Aircraft Co, USA). SME Tech. Paper MR76-874, in, *Proc. Latest Uses of Lasers in Manufacturing Conf.*, 30 November – 2 December 1976, Culver City, CA, 11pp. Society of Manufacturing Engineers, Dearborn, MI, USA,1977.

Non-metal structural materials such as boron epoxy, graphite epoxy, Kevlar, etc., which are difficult to cut by conventional means, can be efficiently cut by lasers into uniquely shaped parts with a high quality edge.

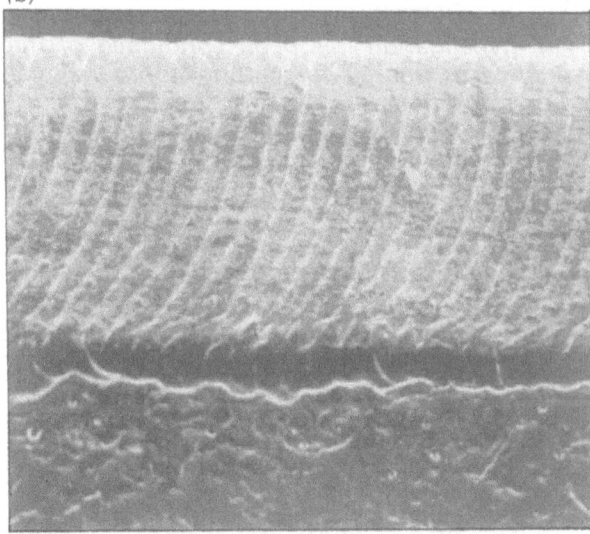

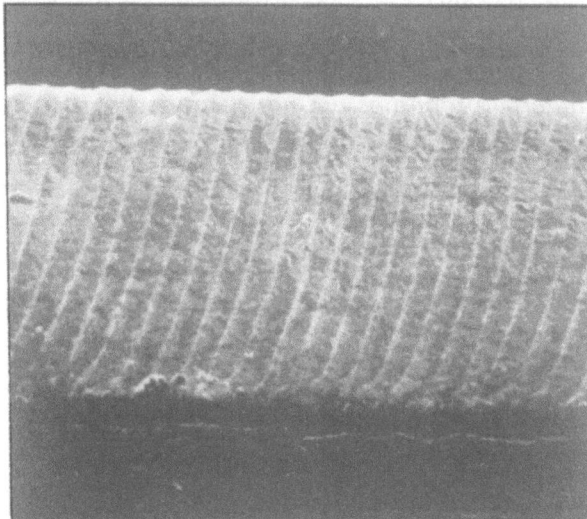

Cut edge quality variation with pulsed length: (a) 1.25ms, (b)1.75ms, (c) 2.0ms, all at 400Hz (See 4.2.2 (49))

4.2.3 (2) Automated system for programed laser cutting of non-metallic materials

Raevich, V. K. et al. *Sov. J. Quantum Electron,* 6(7): 836-838, 1976.

A brief review is given of laser cutting units designed for various purposes. A detailed description is given of an automated system for programmed laser cutting of materials, developed at the All-Union Scientific-Research Institute of Light and Textile Machine Construction. A gas CO_2 laser cutting machine and digital program unit are discussed, and a technical specification of the system is given. (8 refs.)

4.2.3 (3) Laser cutting of kevlar laminates

VanCleave, R. A. (Bendix Corp, USA). Tech. Rept. No. BDX-613-1877, 25pp; September 1977. Bendix Corp, Kansas City, MO, USA.

An investigation has been conducted of the use of laser energy for cutting contours, diameters, and holes in flat and shaped Kevlar 49 fibre-reinforced epoxy laminates as an alternative to conventional machining. The investigation has shown that flat laminates 6.35mm. thick may be cut without back-up by using a high-powered (1000W) continuous wave CO_2 laser at high feed rates. The cut produced was free of the burrs and delaminations resulting from conventional machining methods, and the process cycle time was greatly reduced.

4.2.3 (4) Characteristics of laser cutting kevlar laminates – topical report

Van Cleave, R. A. (Bendix Corp, USA). Tech. Rept. No. BDX-613-2075 (Rev.), 28pp. January 1979. Bendix Corp, Kansas City, MO, USA.

Comparison of laser-cut and machined samples of Kevlar/epoxy laminates has shown that material properties are not degraded by use of the laser. With a 1000W CO_2 laser, the maximum laminate thickness that can be cut satisfactorily is 9.5mm. Power requirements, edge taper, and kerf width for various thicknesses have been established. Partial-depth cuts, controlled and repeatable to within 0.5mm, have been produced in Kevlar/epoxy laminates by scanning the material under a pulsed beam. Fumes produced during cutting require proper exhaust methods to avoid long exposures to benzene, a potential carcinogen.

4.2.3 (5) Paper cutting by CO_2 laser

Hattori, N. et al. (Kyoto University Japan). *Zairyo,* 28(310): 603-609, July 1979. (In Japanese)

Reports on a study in which cutting experiments were carried out on kraft paper and filter paper made of cellulose fibre. The experimental factors were the laser power, defocusing distance, flow rate of assist gas and feed speed. The efficiency of CO_2 laser cutting was evaluated on the basis of external appearance of sections and measurements of kerf width. It was observed that a pyrolysis-induced residue adhered on both cut edges as solid droplets. The minimum power needed to cut filter or kraft paper was about 2.5W when the feed speed was 2m/min.

4.2.3 (6) Economics of cutting wood parts with a laser under optical image analyzer control

Huber, H. A., McMillin, C. W. and Rasher, A. (Michigan State University USA). *Fur. Prod. J.,* 32(3): 16-21, 1982.

Describes use of a laser cutter in a furniture dimension rough mill.

4.2.3 (7) Laser cutting

Gonzalez, J. V. *Deformacion Met.,* (75): 51-57, 1982. (In Spanish)

Illustrates the use of a laser to cut metallic and non-metallic materials. The basic principles of the process are outlined and describes the components of an industrial laser cutting installation. The advantages of the process include speed, minimum cut thickness, minimum heat affected zone and an excellent cut edge finish. Various examples of the use of a laser are given.

4.2.3 (8) Laser cutting of metallic and non-metallic materials with medium powered (1.2kW) CW lasers

Berloffa, E. H. and Witzmann, J. (Voest-Alpine AG, Austria). In, *Proc. SPIE Conf. on Industrial Applications of Laser Technology,* Vol. 398, 19-22 April 1983, Geneva, Switzerland, pp. 354-360. International Society of Optical Engineers, USA, 1983.

Discusses the principle set-up of cutting equipment, the 'Riefen' structure itself, and the specific cut pattern obtained for metallics. Possible instant lens destruction is cited and an explanation is given. (17 refs.)

4.2.3 (9) Laser cutting plastics-final report

Van Cleave, R. A. (Bendix Corp, USA). Rept. no. BDX-613-2906, 20pp; April 1983. (Update of BDX-613-2476, 42 pp; August 1980 and BDX-613-2727, 49pp; December 1981.) Bendix Corp, Kansas City, MO, USA.

Laser cutting and drilling various types of plastic and ceramic materials have been investigated. Nearly 100 different parts currently use laser technology as part of their production processes. A brief discussion of the 1000W CO_2 laser development and production of parts is presented. The 1000W CO_2 laser proved to be a reliable, practical, and cost-effetive machine tool.

4.2.3 (10) Laser material processing of polymers

Nielsen, S. E. (Technical University of Denmark). *Polym. Testing*, 3(4): 303-310, 1983.

The laser processing mechanisms involved in laser drilling, cutting, welding and engraving are described and some typical examples of laser processing are given with attention to processing of polymers.

4.2.3 (11) Reactive gas assisted laser cutting – physical mechanism and technical limitations

Schuocker, D. (University of Technology, Austria). In. *Proc. SPIE Conf. on Industrial Applications of Laser Technology*, Vol. 398, 19-22 April 1983, Geneva, Switzerland, pp. 383-392. International Society of Optial Engineers, USA, 1983. Also in, *Proc. SPIE Conf. on Industrial Applications of High-Power Lasers*, Vol. 455, 26-27 September 1983, pp. 88-95. International Society of Optical Engineers, 1984.

From results obtained with the laser cutting process in acrylic glass of reasonable thickness, it is known that erosion takes place at a nearly vertical plane at the momentary end of the cut. The plane is covered by a molten layer that is heated by absorbed laser radiation and by reaction. The removal of material from that layer is by evaporation and by ejection of molten material due to the friction between the melt and the reactive gas flow. A computer simulation of that model yields a more detailed understanding of laser cutting. (6 refs.)

4.2.3 (12) Cutting of fibre-reinforced polymers with a CW CO_2 laser: an experimental study

Flaum, M. (Foersvarets Forskningsanstalt, Sweden). Rept. No. FOA-C-20527-D9, 53pp; January 1984. (In Swedish)

Describes a CW 250W CO_2 laser cutter, including the gas nozzle with novel features for the protection gas (air, nitrogen, argon, helium, or carbon dioxide). The dependence of maximum cutting speed on different parameters, is reported. A method for cut quality diagnosis is outlined and preliminary tests presented. Good cutting results are noted for aramid fibre/epoxy and glass fibre/epoxy composites, while boron fibre/epoxy and carbon fibre/epoxy were hard to process. Carbon dioxide is preferable for glass fibre, and helium for boron fibre.

4.2.3 (13) Cutting thin sheets of allyl diglycol carbonate (CR-39) with a CW CO_2 laser: instrumentation and parametric investigation

Kukreja, L. M. et al. (Bhabha Atomic Research Centre, India). *Nucl. Instrum. Methods, Phys. Res. Sect A.*, 219(1): 196-198, 1984.

Recent studies have shown that allyl diglycol carbonate, commercially known as CR-39 can detect relativistic oxygen and other heavier nuclei. Describes the use of large sheets of special grade CR-39 (DOP) in Space Shuttle Spacelab-3. As CR-39 is a highly brittle substance, special care is required to cut large sheets. A study of cutting of CR-39 sheets using laser light is described in this paper. A maximum speed up to 200cm/min. is attained.

4.2.3 (14) Damping preform process using laser cutting technology

Pal, G. S. and Snyder, C. G. *IBM Tech. Discl. Bulletin*, 27(5): 2874-2875, 1984.

Describes a process for forming a polymer material to a non-regular shape by laser cutting. The process illustrates the inclusion of a visco-elastic polymer damping material on an irregularly shaped flat plate.

4.2.3 (15) The laser cutting of silicon carbide single crystals

Chepurnov, V. I. et al. *Physics and Chemistry Material Treat.*, 18(2): 134-135, 1984.

Laser radiation can be used for cutting silicon carbide wafers, producing cuts of complex shapes. The decomposition products of silicon carbide do not contain any compounds capable of diffusing into the semiconductor.

4.2.3 (16) Studies on laser cutting of plastic track detector sheets and its effects on track revelation properties

Kukreja, L. M. et al. (Bhabha Atomic Research Centre, India). *Nucl. Tracks Radiat. Meas.* 9(3-4): 199-208, 1984.

A 5-15W CW CO_2 laser is used for cutting with high precision, reproducibility and practically a damage-free edge for the track revelation. Optimum processing parameters and the features of the laser produced cuts are reported. The track revelation sensitivity of some of the plastics, particularly that of the CR-39, enhances due to its laser treatment. (9 refs.)

4.2.4 Laser drilling

4.2.4.1 General

4.2.4.1 (1) Application of ultrasonic laser machining to laser deep-drilling technique

Mori, M. and Kumehara, H. (Gunma University, Japan). *Bulletin Jpn. Soc. Precis. Engng.*, 10(4): 177-178, 1976.

Describes a method in which ultrasonic vibration is exerted on the workpiece being irradiated by a laser beam. Experiments have been conducted in order to study the effects of the ultrasonic vibration on the shape and dimension of holes and the weight of removed materials.

4.2.4.1 (2) Hole drilling with a repetitively pulsed TEA CO_2 laser

Hamilton, D. C. and James, D. J. *J. Phys. D. (Appl. Phys.)*, 9(4): L41-L43, 1976.

4.2.4.1 (3) Production laser hole drilling – now

Battista, A. D. and Shiner, W. H. SME Tech. Paper no. MR76-861, 14pp., 1976. Society of Manufacturing Engineers, Dearborn, MI, USA.

4.2.4.1 (4) Laser drilling

Nagano, Y. *Technocrat*, 11(11): 19-25, 1978.

Lasers applicable to drilling are the ruby, YAG, glass and solid lasers and CO_2 lasers. Properties are discussed and a laser drilling machine is described. In laser drilling, the depth and shape of the hole and the stock removal are affected by the focal length, focal position and energy radiated. Laser machining is capable of various special drillings, one of which is the drilling of a diamond die for wire drawing.

4.2.4.1 (5) Drilling without drills

Bellows, G. and Kohls, J. B. *Am. Mach.*, 126(3): 173-188, 1982.

Several of the following processes are particularly adaptable to economical hole making: electrical discharge machining and wire cutting, electrochemical machining, shaped-tube elec-trolytic machining, electrostream, laser beam machining, electron beam machining, photochemical machining, and ultrasonic machining. Applications and characteristics of each are compared in a chart.

4.2.4.1 (6) Transient photo-acoustic monitoring of pulsed laser drilling

Yeack, C. E. and Melcher, R. L. (IBM Thomas J. Watson Research Center, USA). In, *Proc. 1982 Ultrasonics Symp.*, Vol. 2, 27-29 October 1982, San Diego, CA, (Ed. B. R. McAvoy, pp. 555-558. IEEE, New York, 1982. Also in *Appl. Phys. Letters*, 41(11): 1043-1044, 1982.

4.2.4.1 (7) Investigation of material expulsion mechanism in laser drilling using modelled workpiece

Kato, J. et al. (University of Tokyo, Japan). *Bulletin Jpn. Soc. Precis. Engng.*, 19(2): 133-134, 1985.

In pulsed laser drilling, the material is mainly expulsed in liquid form, and the mechanism of this process is closely related to drilling efficiency, and to the shape and accuracy of the drilled hole. The authors designed a modelled workpiece for laser drilling that is arranged for observation of the phenomena using high speed photography. By this method, the behaviour of the modelled pulsed laser drilling process can be observed in sequence.

4.2.4.2 Laser drilling of metals

4.2.4.2 (1) Influence of laser-supported detonation waves on metal drilling with pulsed CO_2 lasers

Allmen, M. V. and Stumer, E. *J. Appl. Phys.*, 49(11): 5648-5654, 1978.

The drilling process was investigated by measuring the time-resolved laser power reflected specularly from the targets during the interaction and by analysing the craters produced. Target damage was found to be strongly influenced by a laser-supported detonation (LSD) wave in the ambient gas. If the laser fluence exceeded a material-dependent damage threshold, drilling occurred, but the efficiency was inversely related to the duration of the LSD wave. Efficient drilling is possible if the LSD wave can be dissipated within a small fraction of laser pulse duration. Replacing the ambient air by a gas of lower density results in a further significant reduction of LSD wave lifetime, and a correlated increase of the drilling yield.

4.2.4.2 (2) Laser drilling with different pulse shapes

Roos, S. O. *J. Appl. Phys.*, 51(9): 5061-5063, 1980.

Drilling in thin metal plates has been studied with two types of pulses from an ordinary flash-lamp-pumped Nd:YAG laser. The drilling performance with a normal pulse consisting of a 200µs continuous emission preceded by a few relaxation oscillations has been compared with that of a 200µs train of 0.5µs pulses. The pulse train was obtained by introducing a periodic low loss into the laser resonator with the frequency of the relaxation oscillations. It is shown that the use of the pulse train is much superior to the normal pulse for drilling in Al and similar metals.

4.2.4.2 (3) Lasers cast a new light on metalworking

Green, R. G. *Tool. Prod.*, 45(12): 72-83, 1980.

Laser drilling has allowed the hardest materials to be drilled with ease. Extremely high laser cutting speeds have been achieved with Al alloys and stainless steels. Welding with the laser eliminates mechanical distortion and can be precisely controlled. Although the beam must be carefully controlled, lasers offer advantages in heat treating to case harden selective areas of a part.

4.2.4.2 (4) A new method of multiple-shot hole drilling with ruby laser of inconstantly energised pulses

Kumehara, H. and Mori, M. *Bulletin Jpn. Soc. Precis. Engng.*, 14(2): 105-106, 1980.

Energy-inconstancy Dm, Cm and Im pulse-train modes in combination with refocusing length gave trumpet-shape, thick-bottomed, and rear cylindrical holes, in S45C high-C steel specimens. Presents refocusing-length, pulse, and hole-geometry relationships; mechanisms are postulated for the hole-shape dependency on the pulse modes.

4.2.4.2 (5) The effect of F and beam divergence on quality of holes drilled with pulsed Nd:YAG lasers

Bolin, S. R. In, *ICALEO '82, Proc. Materials Processing Symp.*, Vol. 31, 20-23 September 1982, Boston, MA, pp. 135-140. Laser Institute of America, Toledo, OH, USA, 1982.

The increased acceptance and usage of lasers has come from improvements in equipment which have improved drilling performance and have resulted from an acceptance of some of the characteristics of laser holes in areas in which characteristic limitations do not affect engine safety or performance. The most serious of those limitations are taper, diameter variation, drilling depth limits, and microcracking. These limitations, with emphasis on minimising beam divergence are discussed, using Inconel 718 as trial metal alloy.

4.2.4.2 (6) Metal drilling investigation by means of different high-power laser radiation

Cingolani, A., Ferrara, M. and Lugara, M. *Appl. Phys. Comm.*, 2(1-2): 9-16, 1982.

Describes drilling experiments on Cu, Al and brass by Q-switched high-power lasers (Rb, Nd:YAG) and CO_2 lasers. Discusses the experimental results obtained in the intensity range between 10^6 and 10^{10} W/cm^2 in comparison with the theoretical predictions. (8 refs.)

4.2.4.2 (7) Localised metallic melting and hole boring by laser-guided discharges

Gilgenbach, R. M., Horton, L. D. and Ulrich, O. E. *Rev. Sci. Instrum.*, 54(1): 109-113, 1983.

The technique has important applications to a novel method for machining and welding materials, since the melting location can be controlled by adjusting the path of the focused laser beam. These experiments have demonstrated several features of localised metallic melting by laser guided discharges: the melted spot can be scanned by changing only the position of the laser focal spot on the metal sample; the melted spot diameter and profile depend upon the relative timing of the laser pulse and the discharge; defocusing the laser beam has an effect on the melted spot pattern; and hole boring has been accomplished in Al foils which cover the stainless steel electrode. (12 refs.)

4.2.4.2 (8) Acoustic emission monitoring of laser drilling

Clough, R. B., Schaefer, R. J. and Wadley, H. N. G. In, *ASM Metals Congress*, 15-20 September 1984, Detroit, MI, Metals/Materials Technology Series no. 8408-024, 10pp. American Society for Metals, OH, USA, 1984.

Beam power spikes and intervening disintegration products can disrupt or diminish energy transmission to the hole site, resulting in variable hole depth. To improve this situation, a combination of acoustic emission monitoring and laser settings has been investigated to improve hole depth prediction in 2024 Al alloy. The emission is suggested to be generated principally by liquid expulsion; intergranular cracking appears to be a secondary source.

Acoustic emission shows promise as an experimental method for study of directed energy beam-material interactions. (6 refs.)

4.2.4.2 (9) Melting and removing processes of stainless steel by YAG laser irradiation

Watanabe, T. and Shimo, C. *J. Jpn. Soc. Precis. Engng.*, 51(5): 1007-1012, 1985. (In Japanese)

The processes are investigated by measuring the drilled depth and melted depth. The behaviour of the melt is discussed by observing the refleced laser light from the surface of the workpiece and the transmitted laser light through the irradiated part. Results are as follows: melting .and removing processes of the metal are classified into non-melt, melt, removal, and mix types of melt according to the power density. The maximum hole depth closely depends on the power density (W) but not on the irradiation time. The growing of the melted zone and the removal mechanisms of molten material are explained by models based on the experimental results.

4.2.4.3 Laser drilling of non-metals

4.2.4.3 (1) Drilling and cutting of polymethyl metharcrylate (perspex) by CO_2 laser

Berrie, P. G. and Birkett, F. N. (Loughborough University of Technology, UK). *Opt. Lasers Engng.*, 1(2): 107-129, 1980.

Experiments were performed to determine the effect of lens, position and focal plane, speed of cut and power on the cutting and drilling rates of Perspex. The results were compared to thermal conductivity and vapour removal theories. The latter were found to be in good agreement with the experimental observations, and provide a sound basis for the assessment of laser machining of other similar materials.

4.2.4.3 (2) Development of laser drilling techniques to fabricate ceramic distributor plates — phase 1 report

Polk, D. H. (United Technologies Research Center, USA). Rept. No. DOE/ID/12537-1, 62pp; August 1985. United Technologies Research Center, East Hartford, CT, USA.

The objective of this programme was to analytically and experimentally examine the technical and economic feasibility of laser drilling as a means for producing closely spaced arrays of small holes or slots in selected ceramics. Silicon carbide, silicon nitride, 85% alumina, and a SiC fibre composite were all successfully drilled with 3% open area arrays of 0.01in. diameter holes, 0.06in diameter holes, and 0.01 x 0.06in. slots and with no loss of strength except from section area reduction and stress concentrations.

4.2.5 Laser machining

4.2.5.1 General

4.2.5.1 (1) Laser machining takes on practical dimensions in non-contact, localised material processing

Takaoka, T. et al. (Tokyo Shibaura Electronical Co, Japan). *JEE J. Electron. Engng.*, (117): 24-27, 1976.

Shows that when a laser beam is focused through an optical system, the power density at the focal point is enhanced so that almost any substance vapourises instantly or is melted. Laser machining is thus performed through non-contact local heating of materials.

4.2.5.1 (2) Study on ultrasonic laser machining

Mori, M. and Kumehara, H. (Gunma University, Japan). *Ann. CIRP*, 25(1): 115-119, 1976.

Used to improve the accuracy of the shape of laser-drilled holes. Ultrasonic vibration is exerted on a workpiece being irradiated by a laser beam. Experimental results demonstrating the method's efficiency are included.

4.2.5.1 (3) Intensification of laser machining by an electric discharge

Kovalenko, V. S. and Dyatel, V. P. *Elektron. Obrab. Mater.*, (1): 9-11, 1977. (In Russian)

Presents results from a study of laser machining of large diameter holes by introducing additional discharge energy into the zone of action of the laser beam.

4.2.5.1 (4) Micromachining with YAG lasers

Wasko, J. H. SME Tech. Paper no. MR77-984, 15pp; 1977. Society of Manufacturing Engineers, Dearborn, MI, USA.

4.2.5.1 (5) Machine tool design and research

Alexander, J. M. (Ed.) (Imperial College, London, UK). *Proc. 18th Int. Machine Tool Design and Research Conf.*, 14-16 September 1977, London, UK. Imperial College of Science and Technology, London, UK and Macmillan Press Ltd, London, UK, 1978, 860pp.

One hundred papers on advances in metal forming, metal cutting, and machine tool design including laser machining.

4.2.5.1 (6) Applications of high power lasers

Angus, J. C. (Case Institute of Technology, USA). Final Rept. No. NASA-CR-158234, 19pp; March 1979. Case Institute of Technology, Cleveland, OH, USA.

Investigates the use of computer generated, reflection holograms in conjuction with high power lasers for precision machining of metals and ceramics. The reflection holograms meet the primary practical requirement of ruggedness and are relatively economical and simple to fabricate. The technology is sufficiently advanced now so that reflection holography could indeed be used as a practical manufacturing device in certain applications requiring low power densities. However, the present holograms are energy inefficient and much of the laser power is lost in the zero order spot and higher diffraction orders. Improvements of laser machining over conventional methods are discussed and additional applications are listed. Possible uses in the electronics industry include drilling holes in printed circuit boards, making soldered connections, and resistor trimming.

4.2.5.1 (7) An investigation on laser-assisted machining

Capello, G. and Marinsek, G. In, *Int. Colloquium on the Cutting of Metals*, 21-23 November 1979, Saint-Etienne, France, 18pp. Ecole Nationale Superieure des Mines de Saint-Etienne, 1980.

Applying laser machining it is possible to locate the heat source with extreme accuracy, the temperature increase is very fast, the working parameters are easily controllable with a good repeatability and it is possible to heat the material even if the cutting speeds are high due to the high laser power densities available. Outlines some problems in laser-assisted machining such as the low absorption coefficient of metallic materials requiring a surface treatment; operators must be protected behind shields during laser machining; and the cost of a laser installation is expensive.

4.2.5.1 (8) Laser machining

Balbach, J. and Tonshoff, H. K. *VDI-Z*, 122(10): 391-394, 1980. (In German)

The practical applications and basic principles of laser technology and its use in the metalworking field are reviewed. The various types of laser are described, with emphasis on the CO_2 laser. The use of lasers for a variety of operations, including welding, surface treatment and cutting, is discussed and a detailed description of the operation of the laser cutting process is given, with reference to the type of materials applicable, the effect on the boundary zone, the nature of the cut surfaces, processing time and production costs.

4.2.5.1 (9) Application of multi-kilowatt-lasers for the machining of materials

Amende, W. and Zechmeister, H. *VDI-Z.*, 124(15-16): 581-590, 1982. (In German)

The laser beam provides a thermally effective, wear resistant machine tool that can serve in automated production systems for the machining of workpieces. The technical and/or economic advantages offered through its use are compared to other machining methods. The high capacity laser is acquiring an increasing importance, particularly the CO_2 model with a beam performance in the multi-kilowatt range. (8 refs.)

4.2.5.1 (10) Lasers: a machining success story

Walker, R. W. *Photonics Spectra*, 16(9): 65, 66, 68, 70, 72, 1982.

A table of laser machinability of various metals, organic and inorganic materials is provied. A CO_2 laser used with oxygen gas can produce burr-free cuts with cut widths of 0.001in. in 0.12in. thick steels. Discusses techniques to improve capabilities in cutting and welding and the use of lasers in heat treating and surface hardening. Rates for full penetration welds using a CO_2 laser are given.

4.2.5.1 (11) Laser cut blanking tool

Nakagawa, et al. (University of Tokyo, Japan). *Bulletin Jpn. Soc. Precis. Engrs.*, 17(1): 45-46, 1983.

A new type of blanking tool with multi-layered structure was designed and successfully manufactured by using an accurate numerically-controlled laser cutting machine.

4.2.5.1 (12) Laser machining centers spell productivity

Johnson, J. (Control Laser Corp, USA). *Lasers Appl.*, 3(3): 73-77, 1984.

The minicomputer linked to multifunction laser systems has helped the development of laser machining centres with high productivity and cost effectiveness. Virtually any combination of cutting, drilling, welding, heat treating, and scribing/engraving can be performed by a single machine. All operations are performed automatically under computer control, and tool-changing capabilities derive from simple alterations of the laser beam's characteristics to perform drilling, as opposed to welding or cutting, etc.

4.2.5.1 (13) Laser machining in the model shop

Vanderwert, T. L. In, *2nd Biennial Int. Machine Tool Technical Conf. Proc.*, Vol. 4, 5-13 September 1984, McLean, VA, pp. 12/45-56. National Machine Tool Builders Association, McLean, VA, USA, 1984.

Emphasises the primary advantages of laser machining for the model shop as the ability to machine intricate patterns in a wide range of material with short turnaround and low tooling costs. The NC part program that describes the part geometry constitutes the tooling. Time required to generate the program is considerably less than to fabricate or configure conventional tooling, particularly when the laser machining system is interfaced with a CAD system. Furthermore, part programs can be easily and quickly modified for optimisation of part design.

4.2.5.1 (14) Reverse machining via CO_2 laser

Mehta, P., Cooper, E. B. and Otten, R. (General Electric Company, USA). In, *ICALEO '84, Proc. Materials Processing Symp.*, Vol. 44, 12-15 November 1984, Boston, MA, (Ed. J. Mazumder), pp. 168-176. Laser Institute of America, Toledo, OH, USA, 1985.

Consolidation of material via laser has been reported in various forms. This report deals with a special off-shoot of consolidation termed 'reverse machining'. This technology offers the potential for salvaging components which, traditionally, would be classified irrepairable. (10 refs.)

4.2.5.1 (15) YAG laser and its applications

Miura, H., Shiroki, K. -i. and Tsushima, K. (NEC Corp, Japan). In, *ICALEO '84, Proc. of Materials Processing Symp.*, Vol. 44, 12-15 November 1984, Boston, MA, (Ed. J. Mazumder), pp. 269-275. Laser Institute of America, Toledo, OH, USA, 1985.

For the medium-power solid state laser, growing technologies of large size Nd:YAG crystals, Kr arc lamps, and technologies of laser performance were developed. Using a Nd:YAG rod, 8mm. in diameter and 150mm. in length and a prolate type pumping cavity with minor axis of 30cm, an output power of 365W and a conversion efficiency of 3.6% are obtained. In the Tsukuba Test Plant, the 300W laser beam is switched into two optical fibres alternately, and guided to two laser machining heads. The laser beam is used to cut the chips, which grow from the workpiece at the turning process, and to de-burr after mechanical machining of gear.

4.2.5.2 Laser machining of metals

4.2.5.2 (1) Study on laser machining with bleached hologram

Nakahara, S., Fujita, T. and Sugihara, K. *Techno. Rep. Kansai Univ.*, (17): 9-17, 1975.

The machining energy is supplied from the Q-switched 1MW ruby laser. By bleaching, the diffraction efficiency of the hologram rises up to 30%. The samples are thin films, of about 1 mu thickness, of copper, aluminium and nickel deposited on flat glass plates. Multi-spots are machined simultaneously on the sample by a single shot of laser irradiation.

4.2.5.2 (2) Metal machining with a Nd laser

Pippin, C. A. *Manuf. Engng.*, 76(1): 38-41, 1976.

4.2.5.2 (3) Ultra-sonic assisted laser machining

Ebeid, S. J. and Larsson, C. N. (University of Strathclyde, UK). In, *Proc. 18th Int. Machine Tool Design and Research Conf.*, 14-16 September 1977, London, UK, pp. 507-513. Imperial College of Science and Technology, London, UK, and Macmillan Press Ltd, London, UK, 1978.

Studies the effect of low-amplitude ultrasonic vibration on the rate of metal removal during oxygen-assisted CO_2 laser cutting. The distribution of energy during cutting has been assessed using a 400W laser. (4 refs.)

4.2.5.2 (4) Economical sheet-metal working with conventional blanking and laser cutting

Bitzel, H. *Wekstatt Betr.*, 112(9): 686-688, 1979. (In German)

The numerically-controlled laser cutting press, which uses a combination of the conventional blanking process and laser cutting, represents the most economical form of machining of thick sheets for a number of workpieces.

4.2.5.2 (5) Sheet metal profiling with the laser

Greenslade, A. *Sheet Metal Ind.*, 56(5): 430-434, 1979.

Tabulates a selection of materials and their maximum thickness that can be cut with a 500W laser beam. Applications for commercial laser cutting are outlined.

4.2.5.2 (6) Cutting sheet metal with laser beams

Bitzel, H. *Werkstattstechnik*, 70(5): 333-336, 1980. (In German) Also in *Blech Rohre Profile*, 28(8): 369-371, 1981. (In German)

The combined method of punching and laser cutting is the most economical manufacturing method for small and medium sized batches of workpieces with several similar apertures and complex shapes in St3 sheet metal thicknesses up to 4mm. The characteristics of laser machining and the dependency of the cut on the parameters are explained. The construction and the principle of operation of the combined laser cutting machine are described for St 4301.

4.2.5.2 (7) Specific stock removal energies in the laser beam machining of metals and alloys

Dyatel, V. P., Kovalenko, V. S. and Verkhoturov. A. D. *Elektronnaya Obrab. Mater.*, (2): 16-19, 1980. (In Russian)

Laser beam hole piercing tests were carried out to determine the effects of the material and piercing conditions on the specific stock removal energies. These vary with the machining conditions, but no agreement is found between the experimental and theoretical energy values, particularly for Group VI refractory metals and hard alloys.

4.2.5.2 (8) Stainless steel profiling: a review

Hodge, D. *Sheet Metal Ind.*, 57(1): 37-38, 40, 69, 1980.

Cost advantages of air plasma-arc cutting over Ar/H2 plasma are outlined. The cutting speeds on stainless and mild steel of these plasma-arc cutting and conventional gas cutting processes are compared. Various machines including a profiling machine designed for both plasma and laser cutting are described. The two systems effetively cover almost the whole spectrum of stainless steel profiling.

4.2.5.2 (9) Fundamental study on ultrasonic laser machining with CO_2 laser

Kumehara, H. and Mori, M. *Bulletin Jpn. Soc. Precis. Engrs.*, 15(1): 53-54, 1981.

Results for drilling an SK5-steel workpiece ultrasonically vibrated normal to the laser beam axis show a significant decrease of piercing time with increase of the vibration amplitude, accompanied by an increase of the surface transverse section of a through hole. The thickness of the heat affected layer of through holes is nearly uniform, and is greater than that of undrilled holes.

4.2.5.2 (10) Modern alternatives in machining of sheet metal

Geiger, M. and Wissmeier, H. -J. *Werkstattstechnik*, 72(10): 553-562, 1982. (In German)

Describes new cutting methods introduced recently for small- and medium-series production of sheet metal parts. Areas of application include automobile engineering and construction of electrical equipment. Discusses laser machining of Al and steel. (13 refs.)

4.2.5.2 (11) Application of ultrasonic laser machining (ULM) to cutting steel plate

Kumehara, H. and Mori, M. *Bulletin Jpn. Soc. Precis. Engrs.*, 17(1): 9-12, 1983.

Thick carbon-steel plate workpieces were vibrated at 20kHz frequency during the laser beam radiation. Spattering of particles from the cutting point became increasingly violent with increase of the vibration amplitude from nil to 40μm, during which the cut developed from a shallow groove to a clearly incised cut.

4.2.5.2 (12) Laser cuts tough machining problems

Mod. Mach. Shop, 57(7): 78-83, 1984.

Outlines the benefits of laser machining including minimum tooling costs, ease of prototype development, material savings in cutting sheet metal blanks, superior dimensional control

Experimental set-up for the laser machining (See 4.2.5.2 (15))

and repeatability and close nesting, which allow faster delivery of parts to customers. Work at E.R.K. Manufacturing, Inc, is cited.

4.2.5.2 (13) Sheet-metal machining by laser beam cutting

Fehrensen, H. *Werkstattstechnik*, 74(10): 589-591, 1984. (In German)

Lasers are used primarily for cutting and welding and for heat treatment of workpieces in the metal processing industry. Describes the manufacture of sheet metal (St37, stainless steels) workpieces on a laser cutting machine.

4.2.5.2 (14) Energy efficient laser machining

Li, L. J. (Hunan University, People's Republic of China) and Mazumder, J. (University of Illinois, USA). In, *Proc. 2nd Int. Conf. on Lasers in Manufacturing*, 26-28 March 1985, Birmingham, UK, (Ed. M. F. Kimmitt), pp. 23-36. IFS (Publications) Ltd, Bedford, UK, 1985. Also in, *ICALEO '84, Proc. Materials Processing Symp.*, Vol. 44, 12-15 November 1984, Boston, MA, (Ed. J. Mazumder), pp. 177-185. Laser Institute of America, Toledo, OH, USA, 1985.

A new machining process using a laser beam as the tool is proposed and preliminary experiments are carried out to prove the concept. In this process, a laser beam, which is focused by cylindrical lenses to a light strip (band source), is used to melt the micro-area at the interface between the material to be removed and the main body. The chip separated due to melting is removed by a stripping tool during relative motion between the tool and the workpiece. The cutting force is small since the material is removed by micro-area melting as opposed to plastic deformation by a tool for traditional cutting process. The laser cutting tool is free from wear and inertia-related problems. The process is particularly suitable for machining metals such as nickel base alloys which are difficult to cut. Initial data indicates that the finer the focused beam, the better the cutting performance (9 refs.)

4.2.5.2 (15) Process monitoring of high power CO_2 lasers in manufacturing

König, W. et al. (Fraunhofer-Institut für Produktionstechnologie (IPT), West Germany). In, *Proc. 2nd Int. Conf. on Laser in Manufacturing*, 26-28 March 1985, Birmingham, UK, (Ed. M. F. Kimmitt), pp. 129-140. IFS (Publications) Ltd, Bedford, UK, 1985.

Essential parameters affecting the process quality of laser beam machining as imposed by workpiece applications are presented and discussed. The method of transformation hardening as an example illustrates the influence of fluctuations in the beam power profile upon the quality result of this operation.

4.2.5.3 Laser machining of non-metals

4.2.5.3 (1) Manufacturing methods for cutting, machining and drilling composites, volume 1 composites machining handbook

Marx, W. et al. (Grumman Aerospace Corp, USA). Final Tech. Rept. No. AD-B034 148/7, 131pp; August 1978. (Also Vol. 2, Rept No. AD-B034 2022) Grumman Aerospace Corp, Bethpage, NY, USA.

Materials included in this programme included graphite/epoxy and their hybrids, boron/epoxy, kelvar/epoxy and fibreglass/epoxy. Conventional cutting methods were compared to new technology methods such as water-jet, laser and reciprocating cutting.

4.2.5.3 (2) Shaping materials with a continuous wave carbon dioxide laser

Copley, S. M. and Bass, M. (University of Southern California, USA). In, *Applications of Lasers in Material Processing, Proc. of a Conf.*,18-20 April 1979, Washington DC, pp. 121-134. ASM (Materials/Metalworking Technology Series) 1979.

Describes the process in which the laser beam is initially directed along a path that is parallel to the turning axis of a lathe, and it is then reflected from a mirror along a direction perpendicular to the turning axis to a mirror mounted on the cross-slide. After reflection from the cross-slide mirror, it is focused by a ZnSe lens on the workpiece. Two approaches to turning with a laser are laser-assisted machining, where the laser heats a volume of material directly in front of the single point cutting tool; and laser machining, where the laser vapourises the material. Laser machining has been used in shaping silicon compound ceramics, such as Si3N4, SiAlON, and SIC. (6 refs.)

4.2.5.3 (3) Shaping silicon compound ceramics with a continuous wave carbon dioxide laser

Bass, M., Copley, S. M. and Wallace, R. G. (University of Southern California USA). In, *Proc. The Science of Ceramic Machining and Surface Finishing II Conf.*, 13-15 November 1978, Gaithersburg, MD, pp. 283-292. NBS Special Technical Publication 562, National Bureau of Standards, Washington, DC, USA, 1979.

Results are presented on the use of a high-power continuous wave CO_2 laser to shape workpieces of Si3N4, SiA10N and SiC. Several shaping operations have been investigated in the turning configuration including grooving, threading and the production of convex and concave surfaces. Analysis predicts the feed and beam power corresponding to a specific surface roughness grade and effective material removal rate.

4.2.5.3 (4) Laser machining of ceramic (substrates)

Laudel, A. (Bendix Corp, USA). In, *Proc. Technical Program Electro-Optic Laser Conf. and Exposition*, 19-21 November 1980, Boston, MA, pp. 249-254. Industrial and Scientific Conference Management Inc, Chicago, IL, USA, 1980. Also Bendix Corp, Rept, No. BDX-613-2507, CONF-801139-2, 10pp; 1980. Bendix Corp, Kansas City, MO, USA.

Laser machining is used to contour the ceramic substrates and to drill holes in the ceramic for frontside-backside interconnections (vias) and holes for mounting components during the manufacture of hybrid microcircuits. The laser machining process described consistently produces component holes and contours with good surface quality, hole locations, diameter, flatness and metallisation adhesion. The substrates are resistant to repeated thermal shock and are crack-free.

4.2.5.3 (5) Lasers adopt machining center concepts

Bushor, W. E. *Mach. Tool Blue Book*, 76(9): 76-80, 1981.

The automated centre described incorporated pulsed Nd:YAG lasers, a LaserBrain controller minicomputer, a Bridgeport milling base and an 8in. diameter rotary table mounted to the cross-slide of the milling machine base. The centre is designed to drill holes in fibreglass components.

4.2.5.3 (6) Laser hot spot machining

Wallace, R. J. (University of Southern California, USA). Final Rept. No. AD-A150 461/2/XAD, 311pp; March 1983.

Firstly dissertation investigated the shaping of hot pressed S3N4 with a high power CW CO_2 laser. The laser was used to heat the surface of a workpiece forming a groove by vapourisation. Shaping was accomplished by overlapping the grooves. Results are given showing that groove shapes and material removal rates are independent of irradiation environment. This suggests that decomposition is responsible for the removal of material. To further evaluate the potential of this shaping process, 4-point bend specimens were tested with laser machined surfaces. Secondly, a laser machining process for shaping non-solid revolution shapes with a high power CO_2 laser operating in pulse mode was studied and developed. Graphite was used as the model material to develop the process and it was demonstrated that the process can successfully be adapted to machine silicon nitride. Materials were removed a layer at a time, and by controlling the boundaries of the layers a contour shape can be formed.

4.2.5.3 (7) Machining plastics . . . with lasers

Vanderwert, T. L. (Laserdyne Division of Data Card Corp, USA). *Manuf. Engng.*, 91(5): 55-58, 1983.

Points out that lasers can machine plastics to tolerances normally associated with machining and stamping of metals, and discusses basic laser machining principles, economic aspects of the process, and integration of laser beam positioning with robotics/CNC technology.

4.2.5.3 (8) Shaping materials with lasers

Copley, S. et al. (University of Southern California, USA). In, *Materials Processing Theory and Practice, Laser Materials Processing*, Vol. 3 pp. 297-336. North-Holland Publishing Co., Amsterdam, Netherlands, 1983.

Discusses laser turning and milling. Two approaches are considered: laser-assisted machining of metals, in which a laser beam heats material to be machined by a single point cutting tool; and, laser machining for ceramics, in which the laser forms a groove in the material by vapourisation.

4.2.5.3 (9) Laser machining: no longer non-traditional

Krauskopf, B. *Manuf. Engng.*, 93(4): 53-57, 1984.

The CO_2 gas and Nd-doped yttrium aluminium garnet lasers are most frequently used in industrial applications. Lasers can cut and trim metals, composites, plastics, ceramics, paper, rubber, leather and fabrics. No physical contact with the workpiece occurs, which can reduce maintenance equipment and procedures. Numerous applications of lasers are discussed and their capabilities emphasised.

4.2.5.3 (10) A testbed for laser material processing

Kirillin, A. V. et al. (Academy of Science, USSR). *High Temp.*, 22(6): 930-934, 1984.

A testbed is set up for laser machining in a controlled gas medium at temperatures up to 5000K and pressures up to 100MPa. The laser source, the working high-pressure chambers, the beam guiding system, the devices for measuring and adjusting the laser power, the systems for measuring the specimen temperature and monitoring its behaviour, and the gas-vacuum equipment are described.

4.2.5.3 (11) Laser machining, chunk by chunk

Vaccari, J. A. *Am. Mach.*, 129(2): 79, 1985.

Describes the latest approach to machining with lasers, which involves splitting the beam from a laser source into two beams at 90 degrees to each other to remove material in chunks, such as rings in turning operations or rectangular bars in milling operations, instead of vapourising all of the material to be removed. A demonstration model has been built and used to turn acrylic plastic, alumina ceramic, and advanced composites.

4.2.5.3 (12) Trimming trees using a high power CO_2 laser: machining of green and dry wood

Malachowski, M. J. (Cellulose Conversion Enterprises, USA). In, *ICALEO '84, Proc. Materials Processing Symp.*, Vol. 44, 12-15 November 1984, Boston, MA, (Ed. J. Mazumder), pp. 185-192). Laser Institute of America, Toledo, OH, USA, 1985.

To test and evaluate the feasibility of the laser tree trimmer, a PRC LASERblade was used to slice through small branches. Green and dry hardwood and softwood were cut with the beam. Cutting rates varied with the material; depths of cuts were to 15cm. Fire was not a significant problem (14 refs.)

4.3 LASER SURFACE MODIFICATION

4.3.1 General

4.3.1 (1) Effects of process parameters on laser surface modification

Liu, C. A. and Humphries, M. J. (Exxon Research and Engineering Co, USA). In, *ICALEO '83, Proc. Materials Processing Symp.*, Vol. 38, 14-17 November 1983, (Ed. E. A. Metzbower), pp. 108-117. Laser Institute of America, Toledo, OH, USA, 1984.

The results of an experimental study of the effects of various parameters on laser surface modification are discussed. Using 310 stainless steel powder on a type 310 substrate, seven parameters in three general categories were studied: (1) Laser parameters: power density, scan speed, optics, shielding gas. (2) Material parameters: powder size, slurry binder. (3) Alloy modification: boron addition.

Within the experimental ranges studied, the powder size, slurry binder strength, and shielding gas showed little effect on the overall melting efficiency and characteristics, whilst the laser delivey optics and their design showed a profound effect.

Modification of the type 310 SS by the addition of 1.5 weight percent of boron greatly enhanced the melting efficiency and fluidity of the alloy, and reduced the effect of the optic systems. It also resulted in smoother coatings, with less heat effect on the substrate. (8 refs.)

4.3.1 (2) The effects of laser surface melting on the oxidation performance of a Ni-Cr-Al-Hf alloy

Smeggil, J. G., Funkenbusch, A. W. and Bornstein, N. S. (United Technology Research Center, USA). *High Temp. Sci.*, 20(2): 163-182, 1985.

Testing invoived both isothermal and cyclic exposure of the annealed as well as the laser surface processed alloy surfaces to a temperature of 1050°C. For the oxidation of the alloy in the annealed state, complex oxide scales formed that were both adherent and protective. Isothermal and cyclic testing effected a similar hafnium distribution in the oxide scale. Profuse amounts of hafnium-rich oxide prostrusions were found to occur after either the isothermal or the cyclic testing of either annealed or laser-processed surfaces. (11 refs.)

4.3.1 (3) Energy conservation potential of surface modification technologies

Le, H. K., Horne, D. M. and Silberglitt, R. S. (Battelle Pacific Northwest Laboratories, USA). Rept. No. PNL-5538, 131pp; September 1985.

The energy conservation impact is assessed by estimating friction and wear tribological sinks and the subsequent reduction in these sinks when surface modified tools are used. Ion implantation, coatings, and laser and electron beam surface modifications are considered.

4.3.1 (4) Laser magnetic domain refinement

Neiheisel, G. L. (Electro-Optical Instrumentation, Armco Inc, USA). In, *ICALEO '84, Proc. Materials Processing Symp.*, Vol. 44, 12-15 November 1984, Boston, MA, USA, (Ed. J. Mazumder), pp. 102-111. Laser Institute of America, Toledo, OH, USA, 1985.

A high-speed process has been developed that refines the magnetic domain size in transformer steel. This process utilises a CW Nd:YAG laser scanned across the material to cause refinement while not affecting the insulative coating on the steel sheet. The result is a steel that has anywhere from 2 to 14% lower energy loss when used in a core of a transformer. (17 refs.)

4.3.1 (5) The pitting corrosion behaviour of laser surface melted 42∅ and 316 stainless steels

Lamb, M., Steen, W. M. and West, D. R. F. (Imperial College of Science and Technology, London, UK). In, *ICALEO '84, Proc. Materials Processing Symp.*, Vol. 44, 12-15 November 1984, Boston, MA, USA, (Ed. J. Mazumder), pp. 133-139. Laser Institute of America, Toledo, OH, USA, 1985.

The specimens were subjected to slow scan rate polarisation testing in a dilute chloride environment. Results were interpreted in terms of structural segregation and stress effects produced by the laser surface melting. (12 refs.)

4.3.1 (6) Characterisation of surfaces morphology in laser cutting

Querry, M., Yuan, S. F. and Bedrin, C. (Institut National des Sciences Appliquées (INSA), France). In, *Proc. of 3rd Int. Conf. on Lasers in Manufacturing*, 3-5 June 1986, Paris, France, (Ed. A. Quenzer), pp. 55-56. IFS (Publications) Ltd, Bedford, UK, 1986.

The microgeometry of surfaces machined by linear laser cutting (CO_2 source, 500W power, continuous wave; industrial steel work, thickness: 3mm.) is investigated, first in terms of usual roughness criteria (Ra, Rz); then, an algorithm works out the height distributions of the defaults, and calculates the statistical criteria, which characterise their shape, by using Pearson's distributions. The analysis of vertical profile spreading shows that the roughness and its dispersions increase together with the depth under laser beam impact surface; the characteristics of height distributions corroborate and define more precisely this evolution. Finally horizontal spreading of the micro-geometry is analysed and the autocorrelation coefficients of roughness profiles calculated. Influence of the numerical control (NC) system which moves the machined work is clearly shown: the surfaces obtained by using only one NC axis are very different to those resulting from a NC biaxial movement.

4.3.2 Coatings and films

4.3.2 (1) Enhancement of laser machining and aging characteristics of thin metal films by means of a sensitising layer – grain size control

Lou, D. Y. (Bell Laboratories, USA). *J. Appl. Phys.*, 48(5): 2015-2018, 1977.

Reflection losses which occur during the laser machining of thin metal films can be minimised by the deposition of a small grained material between the substrate and the metal film. The paper reports on micro-machining experiments using 30ns Nd:YAG laser pulses at 1.06nm.

4.3.2 (2) Laser enhanced electroplating and maskless pattern generation

von Gutfeld, R. J. et al. (IBM Thomas J. Watson Research Centre, USA). *Appl. Phys. Lett.*, 35(9): 651-653, 1979.

4.3.2 (3) Two-phase mechanism of laser-induced removal of thin absorbing films – 2. Experiment

Metev, S. M. et al. (Leningrad Institute of Fine Mechanics and Optics, USSR). *J. Phys. D. Appl. Phys.*, 13(8): 1571-1575, 1980.

A qualitative and quantitative experimental investigation of laser-induced removal of thin absorbing films has been carried out. The experimental results are compared with the corresponding theoretical investigations, made on the basis of the two-phase removal model. The good agreement between experimental and theoretical results shows that the latter may be used for finding the optimum conditions for laser machining of thin films.

4.3.2 (4) Ceramic coating with high-power CO_2 laser

Ikeda, M. et al. (Electrotechnical Laboratory, Japan). In, *ICALEO '83, Proc. Materials Processing Symp.*, Vol. 38, 14-17 November 1983, Los Angeles, CA, USA, (Ed. E. A. Metzbower), pp. 135-140). Laser Institute of America, Toledo, OH, USA, 1984.

Applications of coated ceramic films having high resistance against wear, heat and oxidation have recently been increasing in various industrial fields. This paper deals with the development of ceramic coating technology using high-power CO_2 laser as a heat source for evaporation of ceramic materials. There have been several reports published on laser evaporation by low-power CO_2 lasers (less than 100W), but few have been reported concerning high-power CO_2 lasers. This method is a kind of vacuum evaporations with a high-power CO_2 laser. The authors have revealed that very hard and firm ceramic films with high adhesion to substrates can be obtained by this high-power laser evaporation technique. This paper describes the deposition rate, hardness and chemical composition of various oxide and nitride ceramic films deposited on Mo substrates.

4.3.2 (5) Laser facing of valve chamfers for internal combustion engines

Andriayakhin, V. M. et al. *Svar. Proizvod.*, (5): 19-20, 1984. (In Russian)

Ni-Cr-B coatings 0.75-0.85mm. thick were applied on chamfers of 40Kh10S2M steel valves by CO_2 laser welding using powder filler. Applied voltage was in the 2.3kV range and the laser gun was moved at 10cm/s. Good service life of these coatings is attributable to fine carbide and boride dispersions in the deposit provided by laser welding.

4.3.2 (6) Laser machining of objects with simultaneous visual monitoring in a copper vapor oscillator-amplifier system

Zemskov, K. I. et al. (Academy of Sciences of the USSR). *Sov. J. Quantum Electron.*, 14(2): 288-290, 1984.

Describes the first experiments on laser micromachining of objects by an amplified light beam of specified configuration with simultaneous visual monitoring of the machining process on a screen. A minimum cut width of 2 mu was achieved when machining metallic films deposited on a dielectric substrate. Measurements of the contrast of the amplified image showed it to vary negligibly, even under conditions of considerable saturation, over a wide range of machining beam powers.

4.3.2 (7) Structure of the surface in laser cutting fibrous cellulose materials with a polymer coating

Grigor'ev, V. P., Gudkov, V. K. and Mirkin, L. I. *Phys. Chem. Mater. Treat.*, 19(1): 11-12, 1985.

Mechanical and laser cutting of paper with a polymer coating based on titania is compared. Guillotining is accompanied by delamination and crumbling of a large part of the polymer coating, whereas laser cutting results in a straight and solid edge due to the effect of 'welding' the coating to the fibrous cellulose base.

4.3.2 (8) Surface structure studies of chromium-tin alloys fabricated by ion implantation and laser doping

Babikova, Yu. F. et al. (Institute of Engineering and Physics, USSR). *Sov. Phys. Tech. Phys.*, 30(2): 257-259, 1985.

Ion implantation and laser doping are promising tools for the fabrication of films with specified physical, chemical, and mechanical properties on metal and alloy surfaces. Laser doping involves irradiating a metal coated by a layer of the doping element. The laser beam melts the surface layer of the metal, the elements are mixed in the liquid phase, and the melt crystallises rapidly. Analyses and compares the surface structures of chromium-tin alloys fabricated by ion implantation and laser doping of chromium using the stable 119Sn isotope.

4.3.2 (9) CO_2 laser deposition of a cobalt base alloy on a 12% chromium steel

Com-Nougué, J. and Kerrand, E. (Laboratoires de Marcoussis (CGE) France). In, *Proc. 3rd Int. Conf. on Lasers in Manufacturing*, 3-5 June 1986, Paris, France, (Ed. A. Quenzer), pp. 191-196. IFS (Publications) Ltd, Bedford, UK, 1986.

A laser coating process has been set up and its performances evaluated in the case of a cobalt base alloy to be deposited on a chromium steel. The process is based on the injection of a pre-alloyed powder in the partially focused beam near the laser impact on the workpiece. The experiments were carried out at a laser power less than 3kW with traverse speeds 20 to 70 cm/min. Some surfaces were coated by performing successive passes with overlapping rates at 20 to 50%. The coatings so obtained are characterised by good compactness and a very fine structure free from defects such as porosity. In addition, a very low dilution of the steel substrate is observed. Thicknesses up to 1.36mm were achieved in these conditions.

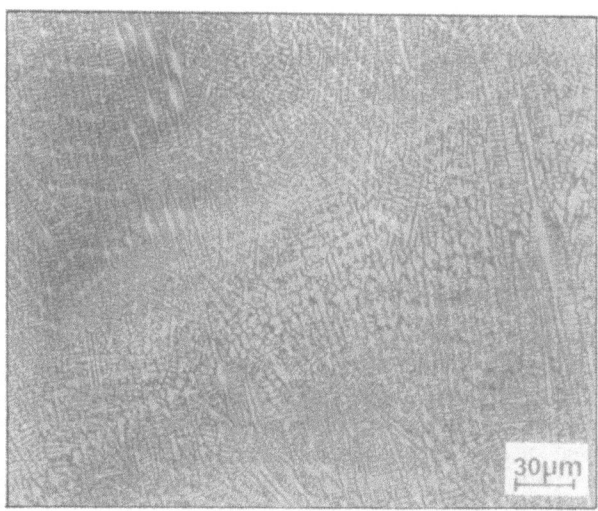

Structure of the coating observed in the overlapping zone of two successive passes (See 4.3.2 (9))

4.3.3 Cladding and alloying

4.3.3 (1) Laser cladding by powder injection

Weerasinghe, V. and Steen, W. M. (Imperial College, UK). In, *Proc. 1st Int. Conf. on Lasers in Manufacturing*, 1-3 November 1983, Brighton, UK, (Ed. M. F. Kimmitt), pp. 125-132. IFS (Publications) Ltd, Bedford, UK, 1985.

A process of laser cladding, using argon-borne powder injection, is described. The effect of the process parameters on the quality of the clad layer is discussed. Process data for flat plate cladding of stainless steel 316 to mild steel EN3 are presented. The significant effect of recycling the reflected energy on the cladding rate is reported.

Laser cladding by powder injection (See 4.3.3 (1))

4.3.3 (2) Laser hardsurfacing of gas turbine blade shroud interlocks

Macintyre, R. M. (Rolls-Royce Ltd, UK). In, *Proc. 1st Int. Conf. on Lasers in Manufacturing*, 1-3 November 1983, Brighton, UK, (Ed. M. F. Kimmitt), pp. 253-261. IFS (Publications) Ltd, Bedford, UK, 1983.

Describes the application of a cobalt based hardsurfacing alloy to a nickel based gas turbine blade for aero-engine use. The prior technique, manual argon arc welding was time-consuming, critically dependent upon operator skill and suffered from a lack of reproducibility, particularly in the level of nickel dilution produced. The use of a laser and a blown powder technique to introduce the hardsurfacing alloy enables clad deposits of optimum wear characteristics, free from heat affected zone cracking and of a consistent quality to be produced. The process is able to be automated and run under microprocessor control.

4.3.3 (3) Lasers effective for attaching Cu-Ni cladding to ship steel

Banas, C. M. (United Technologies Research Center, USA). *Weld. Met. Fabr.*, 51(6): 297-299, 1983.

Penetration of approximately 20% beyond the clad thickness provides a good bond and minimal cracking. Dilution of the surface material by subtrate steel can be held to about 10%, which is below the level for onset of corrosion problems. Laser seam welds can be readily formed in Cu-Ni alloy at high speed. The seam weld can also serve as a means of attaching the cladding.

4.3.3. (4) Cladding of austenitic stainless steel by laser. Poster N. 6.9

Cantello, M. et al. (Union Internationale d'Electrothermie, France). Rept. no. CONF-8406252-97, 14pp. 1984. Presented at 10th UIE Congress on Electroheat for Improved Economy, 18th June 1984, Stockholm Sweden.

Study of the hard facing of austenitic AISI 304 stainless steel by SF6 Stellite. The effective power and translation speed values of the specimen observed under a laser beam ranged respectively from 9 to 11 kW and from 0.125 to 1m/min.

4.3.3 (5) Advances in laser cladding process technology

Eboo, G. M. and Lindemanis, A. E. (Quantum Laser Corp, USA). In, *Proc. SPIE Applications of High Power-Lasers*, Vol. 527, 22-23 January 1985, Los Angeles, CA, pp. 86-94.

The effective application of laser cladding to industrial components has required significant improvements in the process technology, particularly in the techniques for depositing the hardfacing alloy while manipulating the component and laser beam. Reviews advances in equipment, such as dynamic powder feeders, together with the effects of process parameters on cladding quality. (12 refs.)

4.3.3 (6) Laser cladding with mixed powder feed

Takeda, T., Steen, W. M. and West, D. R. F. (Imperial College, UK). In, *ICALEO '84, Proc. Materials Processing Symp.*, Vol. 44, 12-15 November 1984, Boston, MA. (Ed. J. Mazumder), pp. 151-158. Laser Institute of America, Toledo, OH, USA, 1985. Also in, *Proc. 2nd Int. Conf. on Lasers in Manufacturing*, 26-28 March 1985, Birmingham, UK, (Ed. M. F. Kimmitt), pp. 85-96. IFS (Publications) Ltd, Bedford, UK, 1985.

Cladding a metallic surface by blowing powder into a melt pool generated by a laser has been shown to be possible in previous work. In this paper the process is developed further in three ways to allow in situ alloying by blowing mixed powder into the laser generated melt pool. In the first method a single hopper containing a mixed powder feeds a single argon blown delivery pipe. In the second method a multiple hopper system feeds the individual alloying elements into a single argon blown delivery pipe. In the third method a multiple hopper system feeds a multiple delivery pipe arrangement.

It is shown in these preliminary results with the Fe/Cr/Ni system that in situ alloying is possible though homogeneity is dependent upon the cladding speed, location within the tracks and impingement location of the powder feed. (8 refs.)

4.3.3 (7) Laser cladding techniques for application to wear and corrosion resistant coatings

Koshy, P. (Joy Manufacturing Co, USA). In, *Proc. SPIE Applications of High Power Lasers*, Vol. 527, 22-23 January 1985, Los Angeles, CA, pp. 80-85, 88.

Laser cladding techniques and equipment were developed to apply wear/corrosion-resistant alloys on valve sealing surfaces. The new techniques allow cladding on flat, cylindrical and contoured surfaces more economically than prior processes and techniques used. By using the laser cladding technique for production parts advantages are less than 2% dilution of base metal; improved surface wear and corrosion properties, reduced cracking suscepti-

bility, minimal warpage and distortion of clad components; high integrity metallurgical bond; suitability for full automation; considerable savings of expensive cladding materials due to the consistency and uniformity of applied cladding.

4.3.3 (8) Laser surface alloyed Fe-Cr-C

Das, S. et al. (University of Illinois, USA). In, *Proc. 2nd Int. Conf. on Lasers in Manufacturing*, 26-28 March 1985, Birmingham, UK, (Ed. M. F. Kimmitt), pp. 73-84. IFS (Publications) Ltd, Bedford, UK, 1985.

Surface-related failures such as corrosion and wear can be remedied by modifying the surface chemistry by laser surface alloying. Using this process a wide range of alloys can be generated with novel microstructure. Microstructure and composition of laser surface alloyed Fe-Cr-C alloy is characterized by TEM, scanning Auger microprobe, SIMS, and Microprobe. Highly refined microstructure with metastable crystalline, and amorphous phases are observed. Both Cr and C are found to be distributed uniformly. Impurities such as P, S, Si, etc., seems to be responsible for amorphous phases. Due to rapid solidification even high chromium content alloys often retained F.C.C. structure. Chromium carbides are observed inside the grain. Their microstructure and their implications are discussed.

4.3.3 (9) Laser surface alloying

Semiletova, E. F. and Dumbadze, T. H. (Georgian Polytechnic Institute, USSR). In, *Proc. 2nd Int. Conf. on Lasers in Manufacturing*, 26-28 March 1985, Birmingham, UK, (Ed. M. F. Kimmitt), pp. 97-108. IFS (Publications) Ltd, Bedford, UK, 1985.

Outlines the results of investigation of surface alloying of carbon materials with WC, WC + Co type powders by the effect of impulse and continuous laser emission. it has been shown that the process of alloying and the quality of irradiated surface change depending on the method of powder supply to the heating zone and the type of emission used.

4.3.3 (10) Structure and corrosion resistance of AN Mg-Li base alloy after laser treatment

Kalimullin, R. K. and Kozhevnikov, Y. Y. *Met. Sci. Heat Treat.*, 27(3/4): 272-274, 1985.

Laser radiation surface treatment of MA21 alloy increases corrosion resistance by 30 times with pulsed treatment or by more than an order of magnitude with continuous treatment. In the laser treatment zone MA21 acquires a finely dispersed structure with an average grain size of up to 2μm.

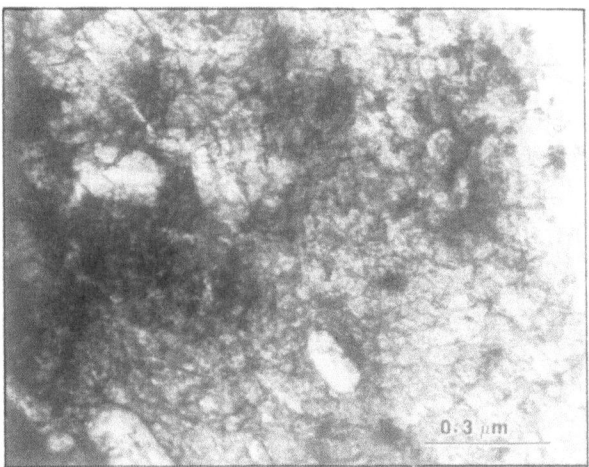

An amorphous region in the Fe-Cr-C alloy (See 4.3.3 (8))

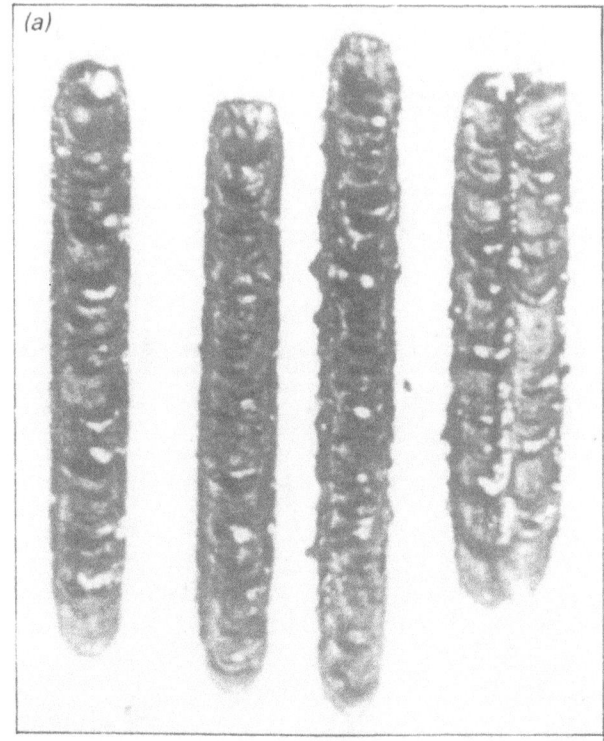

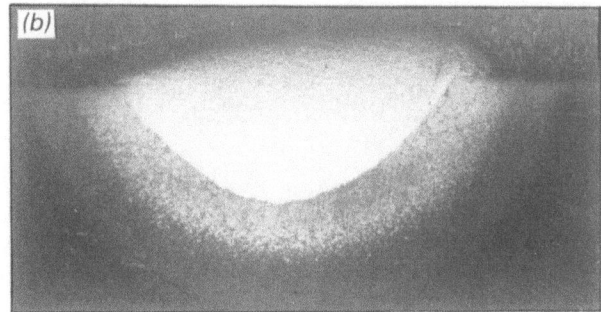

Alloying zones obtained with CO_2 laser emission: (a) view from above, (b) cross-section (See 4.3.3 (9))

4.3.3 (11) Wear properties of laser alloyed and clad Fe-Cr-Mn-C alloys

Eiholzer, E., Cusano, C. and Mazumder, J. (University of Illinois, USA). In, *ICALEO '84, Proc. Materials Processing Symp.*, Vol. 44, 12-15 November 1984, Boston, MA, (Ed. J. Mazumder), pp. 159-167. Laser Institute of America, Toledo, OH, USA, 1985.

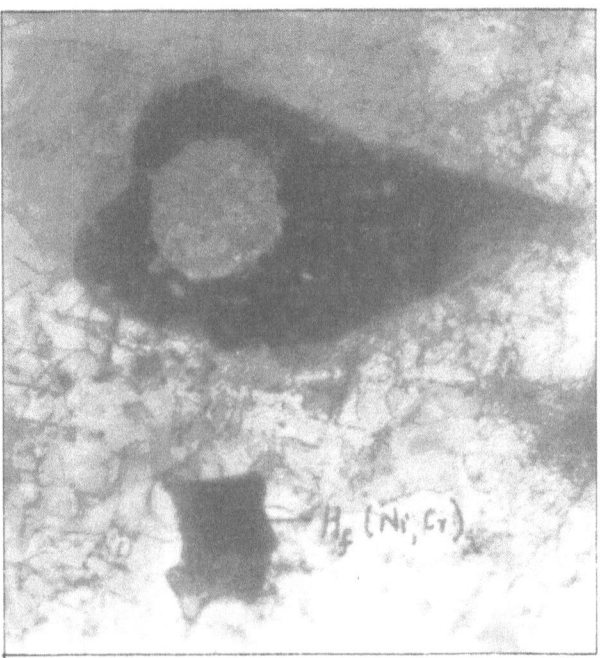

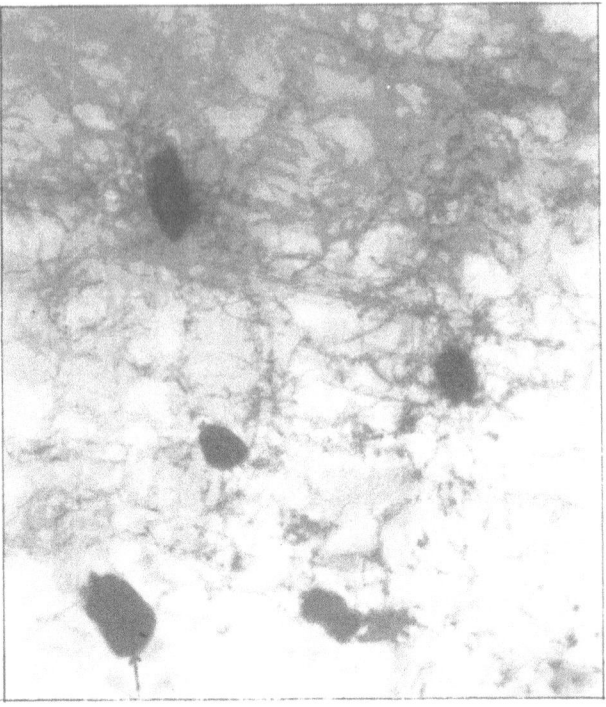

General survey of TEM sample using the undissolved Hf particles, Hf rich precipitates and high density of dislocations in the matrix of laser clad alloy (See 4.3.3 (12))

The abrasive wear resistance of a Fe-Cr-Mn-C clad, produced by laser cladding techniques, was studied for a possible alternative to Co-alloys used in hardfacing for wear applications. The clad was produced by using a pneumatic powder delivery system to inject a Cr-Mn-C powder mixture into a molten pool produced in the AISI 1016 steel substrate by a 10kW CW CO_2 laser. The molten pool area was shielded with argon gas to suppress vapourisation of the Mn, resulting in a non-porous clad. This preliminary study has shown that wear resistance can be significantly improved by laser alloying and cladding of Cr-Mn-C alloy to a mild steel substrate. The laser processed material composed of Fe-Cr-Mn-C exhibited far superior wear properties than Stellite 6 during block-on-cylinder tests. The wear resistance is attributed to dendrite structures and chrome carbides. The finer the dendrites, the higher the wear resistance. Using a two level factorial experimental design matrix gave insight as to which laser process variables result in better wear resistance. The combination of low laser power and high table velocities produced better wear-resistant materials for laser surface alloying.

4.3.3 (12) In situ formation of Ni-Cr-Al-R.E. alloy by laser cladding with mixed powder feed

Singh, J. and Mazumder, J. (University of Illinois, USA). In, *Proc. 3rd Int. Conf. on Lasers in Manufacturing*, 3-5 June 1986, Paris, France. (Ed. A. Quenzer), pp. 169-180. IFS (Publications) Ltd, Bedford, UK, 1986.

The oxidation-resistant materials for service at elevated temperatures must satisfy two requirements: diffusion through oxide scale must occur at the lowest possible rate and oxide scale must resist spallation. The formation of an Al_2O_3 protective scale fulfills the former requirements but its adherence is poor. Rare earths such as Y or Hf are added to improve adhesion. In this paper, in situ Ni-Cr-Al-Hf alloy has been developed by laser surface cladding with mixed powder feed. A 10kW CO_2 laser was used for laser cladding. Optical, SEM, and STEM microanalysis techniques were employed to characterise the different phases produced during laser cladding process. Microstructural studies showed a high degree of grain refinement, increased solid solubility of Hf in matrix and Hf rich precipitates. Deals with the microstructural development during laser cladding process.

4.3.3 (13) Laser cladding and alloying

Coquerelle, G., Collin, M. and Fachinetti, J. L. (Etablissement Technique Central de l'Armement (ETCA), France). In, *Proc. 3rd Int. Conf. on*

Lasers in Manufacturing, 3-5 June 1986, Paris, France. (Ed. A. Quenzer), pp. 197-206. IFS (Publications) Ltd, Bedford, UK, 1986.

Surfaces of materials can be modified by laser alloying and cladding processes to improve corrosion, fatigue and wear resistance. Several processes can be used including chemical surface modification by gaseous reaction and injection of particles. New surface materials are elaborated by alloying or cladding with or without bulk reaction. Hardness and wear resistance of titanium alloys can be enhanced by gaseous nitriding and carburising processes. Injection of silicon particles in aluminium alloys increases wear properties and decreases friction coefficient with graphite addition. In the cladding process, corrosion and wear of iron alloys and superalloys can be decreased by using stainless, stellite and ceramic powders.

4.3.3 (14) Laser surface cladding and residual stresses

Hernandez, J. et al. (INSA/CALFEMAT/GEMPPM, France). In, *Proc. 3rd Int. Conf. on Lasers in Manufacturing*, 3-5 June 1986, Paris, France, (Ed. A. Quenzer), pp. 180-190. IFS (Publications) Ltd, Bedford, UK, 1986.

The laser can be used to fuse a surface layer on to a substrate. This process can be performed in a variety of ways, in particular by blown powder fusion without preheating. In this case, a minimal dilution and distortion, a very high quality clad structure both in the bulk of the clad and in the interface zone are obtained. A specific method has been used to determine the residual stresses field induced by this treatment in the case of a stellite clad and a nickel-base hardfacing alloy.

The residual stresses distributions which have been measured are very interesting in comparison with the stresses obtained by plasma spray. The metallurgy of the clad layer is discussed to understand these distributions.

4.3.3 (15) Properties of laser gas alloyed titanium

Bergmann, H. W. and Lee, S. (Technische Universität Clausthal, W. Germany), and Bell, T. (Wolfson Institute for Surface Engineering, UK). In, *Proc. 3rd Int. Conf. on Lasers in Manufacturing*, 3-5 June 1986, Paris, France, (Ed. A. Quenzer), pp. 221-232. IFS (Publications) Ltd, Bedford, UK, 1986.

Gas alloying of titanium components is possible during laser melting if a suitable reactive gas like N_2 is present. Such treatments enable the production of 0.3 – 1mm thick, hard and wear-resistant surface layers, depending on the pro-

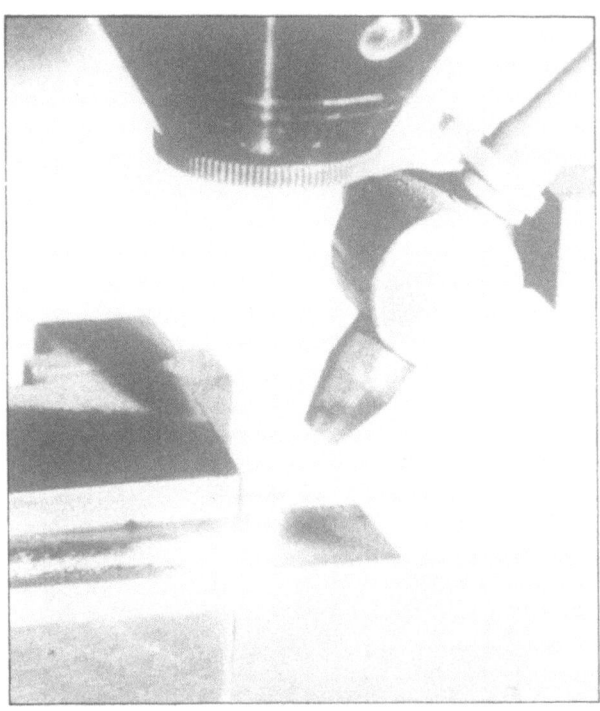

General view of powder coating (See 4.3.3 (13))

Typical microstructure of laser gas nitrided TiAl6V4 obtained at short interaction times and increasing magnification (See 4.3.3 (15))

High power CO₂ laser used in surface treatment tests (See 4.3.3 (17))

4.3.3 (17) Surface treatments by high power CO₂ laser: resolidification of plasma-sprayed coatings and laser cladding

Be, C. A. and Cerri, W. (CISE SpA, Italy) and Magrini, M. and Ramous, E. (Università di Padova, Italy). In, *Proc. 3rd Int. Conf. on Lasers in Manufacturing*, 3-5 June 1986, Paris, France, (Ed. A. Quenzer), pp. 215-220. IFS (Publications) Ltd, Bedford, UK, 1986.

Surface treatments tests have been performed at CISE using a high power CO₂ laser designed and realised by CISE.

Two different techniques have been considered: laser resolidification of plasma sprayed coatings; laser cladding by blowing a powdered clad material into the melt pool generated by the laser on the substrate surface.

Laser consolidation eliminates the chemical and structural defects of plasma sprayed coatings (poor adhesion, porosities, inhomogeneous structure) resulting in an improvement of the cladding corrosion properties. Laser cladding by powder injection produces homogeneous and uniform clad layer, with a metallurgical bonding to the substrate and a low dilution of the base metal. Laser consolidation and cladding of nickel base alloy on stainless steel are reported.

cess parameters. The influence of feed rate and laser power on the nitriding behaviour and the resulting properties are reported.

4.3.3 (16) Silicon surface alloying of steel by laser treatment

Ramous, E. et al. (Università di Padova, Italy). In, *Proc. 3rd Int. Conf. on Lasers in Manufacturing*, 3-5 June 1986, Paris, France, (Ed. A. Quenzer), pp. 207-214. IFS (Publications) Ltd, Bedford, UK, 1986.

Observations on the structure of a laser silicon surface-alloyed low alloy steel are reported. A CO₂ (AVCO) laser (high-power laser 15kW) was used, with a silicon powder deposited on to the steel substrate. The treated layers were characterised by standard metallographic methods, microhardness measurements, SEM, EPM and X-ray diffraction.

According to different laser parameters, melted layers in the range 0-500μm were obtained. Silicon alloying reveals differentiated distributions in the layer, depending on laser conditions. Silicon significantly affects the hardness of the treated surface, mainly through the modification both of the ratio ferrite/martensite and the martensite carbon content.

4.3.4 Heat treatment and surface hardening

4.3.4 (1) Focus is on lasers for fabricating metal products

Miska, K. H. *Mater, Engng.*, 86(5): 50-52, 1977.

Brief papers emphasising the value of the laser in welding and heat treatment, drawing attention to the advantages particularly in heat treatment.

4.3.4 (2) Metals handbook volume 4: Heat treating

American Society for Metals, OH, USA, 1981, 826pp., 9th edition.

Provides information on the heat treating and case hardening of steel, heat processing equipment, furnace control instrumention, furnace atmospheres and carbon control, localised heat treating, the heat processing of powder metallurgy parts and the heat treating of cast irons, tool steels, stainless steels, heat resisting alloys and non-ferrous metals. Includes coverage of fluidised bed processes, vacuum carburising, microprocessors, and laser and electron beam heat treating.

4.3.4 (3) Laser heat treatment

Gregson, V. G. (Coherent Inc, USA). In, *Materials Processing Theory Practice, Vol. 3, Laser Materials Processing*, pp. 201-233. North-Holland Publishing Co, Amsterdam, Netherlands, 1983.

Laser heat treatment is primarily used on steels and cast irons with sufficient carbon content to allow hardening. The metal surface is first prepared with an absorbing coating. After the coating is applied, the laser beam is directed to the surface. A simple thermal model of one-dimensional heating and cooling can be used to calculate the transient temperatures at the surface and at distances below the metal surface. The model can be extended by using simple or complex models of diffusion which add details of the metallurgical response. Metal heating and cooling, metallurgy, and hardening mechanisms are discussed.

4.3.4 (4) Laser machining and laser heat treatment

Megaw, J.H.P.C. (UKAEA Culham Laboratory, UK). In, *Proc. IEE Colloquium on New Tools for Industrial Processes*, (Digest No. 37), 14th April 1983, London, UK, 4pp. Institution of Electrical Engineers, London, UK, 1983.

Medium- and high-power CO_2 laser technology is still attracting much interest and even winning a modest growth in its adoption. No doubt this is due to a heightened awareness of the need for increased productivity and efficiency. The capital costs and technical sophistication are still high, and the time, effort and care required to transfer technology from laboratory to production line should not be underestimated.

4.3.4 (5) Laser surface melt hardening of S. G. irons

Hawkes, I. C. et al. (Imperial College, UK). In, *Proc. 1st Int. Conf. on Lasers in Manufacturing*, 1-3 November 1983, Brighton, UK, (Ed. M. F. Kimmitt), pp. 97-108. IFS (Publications) Ltd, Bedford, UK, 1983.

A Control Laser Ltd 2kW FAF CO_2 laser has been used to surface melt line traces on fernitic and perlitic S. G. iron samples, using a 1.5kW beam of 0.4 – 3.0mm diameter traversed at 5-400mms^{-1}. The effect of these process variables on melt profile, microstructure, hardness, surface finish and solidification cracking are considered.

4.3.4 (6) Opportunities for laser processing in automotive manufacturing

Parsons, G. H. (BL Technology Ltd, UK). In, *Proc. 1st Int. Conf. on Lasers in Manufacturing*, 1-3 November 1983, Brighton, UK, (Ed. M. F. Kimmitt), pp. 117-124. IFS (Publications) Ltd, Bedford, UK, 1983.

Describes the initiation of BL Technology's laser development programme, its scope and objectives together with a brief description of the laser laboratory equipment and layout. There follows a discussion of the role of laser transformation hardening together with the results of a series of hardening trials on typical car industry steels and irons. Applications of laser hardening in the form of selected component case studies are given. The paper continues with examples of cutting applications, and finally concludes with observations on the prospects for volume production implementation of the processes described.

4.3.4 (7) Remelting surface hardening of cast iron by CO_2 laser

Ricciardi, G. et al. (University of Padua, Italy). In, *Proc. 1st Int. Conf. on Lasers in Manufacturing*, 1-3 November 1983, Brighton, UK, (Ed. M. F. Kimmitt), pp. 87-95. IFS (Publications) Ltd, Bedford, UK, 1983.

With this method it is possible to obtain a layer of ledeburitic high hardened structures. The use of high-power CO_2 lasers in surface remelting hardening is justified by high flexibility and the possibility of treating complex shapes. The author treated perlitic grey cast iron by two different models of high-power CO_2 lasers (2.5kW and 15kW). Tests were carried out to investigate the cracks and porosity formation related to laser beam power, specimen speed, shielding gas and use of preheating. Tribological tests were carried out to compare machine parameters with wear resistance of the treated materials.

Spiral hardening of crankshaft pins (See 4.3.4 (6))

4.3.4 (8) Condition setting method utilising data base system in CO₂ laser surface hardening

Arata, Y., Inoue, K. and Matsumura, S. (Osaka University, Japan). In, *ICALEO '83, Proc. Materials Processing Symp.*, Vol. 38, 14-17 November 1983, Los Angeles, CA, (Ed. E. A. Metzbower), pp. 100-107. Laser Institute of America, Toledo, OH, USA, 1984.

The effective method of investigating the interaction model in heat processing is discussed. The computer processing system in this method consists of the data base and several peripheral modules. The hardness estimation of CO₂ laser surface hardening is carried out as the concrete application of this system. The automatic evaluation of the hardness estimation models is discussed. These models are implemented by composing and utilising small and simple program modules which are evaluated automatically and improved by steps. This process is very effectively executed in the whole system. The most suitable model that has been evaluated by the above method can be applied to the hardness estimation within the wide range of condition.

For verification, the automatic condition setting is used. The estimated results of the set condition closely correspond to the experimental surface hardened results.

4.3.4 (9) Eddy current monitoring of laser-heated metals: a means of process control?

Vanderwert, T. L. (Laserdyne Corp, USA). *Laser Focus/Electro-Opt.*, 20(4): 52-56, 1984.

4.3.4 (10) Effect of kinetics of transformation limited by diffusion, in calculating the hardened layer thickess of laser heat treatment of steel

Gurvich, L. O. and Sobol, E. N. *Russ. Metall. Met.*, (6): 149-153, 1984.

Considers a procedure that permits more realistic consideration of the heat and diffusion sides of the (alpha yields gamma) transformation process on laser hardening of steel, thus considerably refining the hardened layer thickness calculation. The proposed approach analyses the dependence of the hardened layer thickness on the laser beam size and capacity, the testpiece movement velocity, the thermophysical properties and chemical composition of the steel, and also the state of the initial steel structure.

4.3.4 (11) Laser transformation hardening of a medium carbon steel

Oakley, P. J. (Welding Institute, UK). In, *ICALEO '83, Proc. Materials Processing Symp.*, Vol. 38, 14-17 November 1983, Los Angeles, CA, (Ed. E. A. Metzbower), pp. 118-126. Laser Institute of America, Toledo, OH, USA, 1984.

The results of laser transformation hardening trials on a medium-carbon steel are described. A 2kW fast axial flow laser was used, producing a defocused beam with a conventional plano-convex lens and also an annular beam with an Axicon lens. Assessment of the hardened zones was made by examining the morphology and microstructure. Trends in the morphology of the hardened regions were predictable, but a variation in microstructure was observed that could not be related to the laser process parameters. (5 refs.)

4.3.4 (12) Surface hardening of titanium by laser nitriding

Katayama, S. et al. (Osaka University and Nissan Motor Co. Ltd, Japan). In, *ICALEO '83, Proc. Materials Processing Symp.*, Vol. 38, 14-17 November 1983. Los Angeles, CA, (Ed. E. A. Metzbower), pp. 127-134. Laser Institute of America, Toledo, OH, USA, 1984.

Nitriding and related hardening of titanium (Ti) and its alloys were investigated as a new method of laser surface treatment by using a pulsed Nd:YAG laser of $1.06\mu m$ wavelength with 3.6ms pulse width. When a laser beam was irradiated on Ti plate in nitrogen (N) atmosphere, it was found that a flat and smooth surface covered with TiN nitride layer could be produced by selecting a good combination of laser irradiation conditions such as beam energy, power density and the distance from the lens focal point to the material surface. Compared with Ti base metal having a Vickers hardness (H_v) of about 200, the hardness of the laser-nitriding surface could exceed about 700 H_v at a single pulse shot and increase up to 1700H_v at overlap of twenty shots; also, the fusion zone could be hardened up to about 400 to 700 H_v. These hardening mechanisms were interpreted in terms of the formation of TiN nitride layer on the surface and the enrichment of nitrogen in the fusion zone. Nitriding and hardening phenomena were observed as well in the case of titanium alloys such as Ti-6Al-4V and Ti-6Al-6V-2Sn. Moreover, it was confirmed that nitriding and hardening of Ti and its alloys were feasible by the use of a CW CO₂ laser. (6 refs.)

4.3.4 (13) Controlled surface temperature laser heat treating

De Pascale, O. and Esposito, C. (Centro Laser, Italy). *Opt. & Lasers Engng.*, 6(3): 157-163, 1985.

Key features are the ease of predicting working results and the capability of obtaining on-line process diagnostics for use in an automated industrial environment.

4.3.4 (14) Development of a numerical code to describe heat diffusion in view of its applicability to heat treatment of metals with continuous CO_2 laser irradiation.

Stern, G. (Institut Franco-Allemand de Recherches, France). Rept. No. ISL-CO-207/85, 34 pp; May 1985. (Text in French, summary in German) Institut Franco-Allemand de Recherches, Saint-Louis, France.

The model used confines itself to a calculation of the heat diffusion in a two-dimensional, axisymmetrical configuration. It takes the change in absorption for the wavelength of the CO_2 laser, of the specific heat and the heat conductivity as a function of the temperature into account. The boundary conditions take heat loss through radiation and convection into account, which are attributable to the use of an argon flow so as to avoid oxidation.

4.3.4 (15) Heat treatment of steels using different high power CO_2 laser beam intensity distributions

Cerri, W. and Vendramini, A. (CISE SpA, Italy), and Ramous, E. (Università di Padova, Italy). In, *Proc. 2nd Int. Conf. on Lasers in Manufacturing*, 26-28 March 1985, Birmingham, UK, (Ed. M. F. Kimmitt), pp. 301-308. IFS (Publications) Ltd, Bedford, UK. 1985.

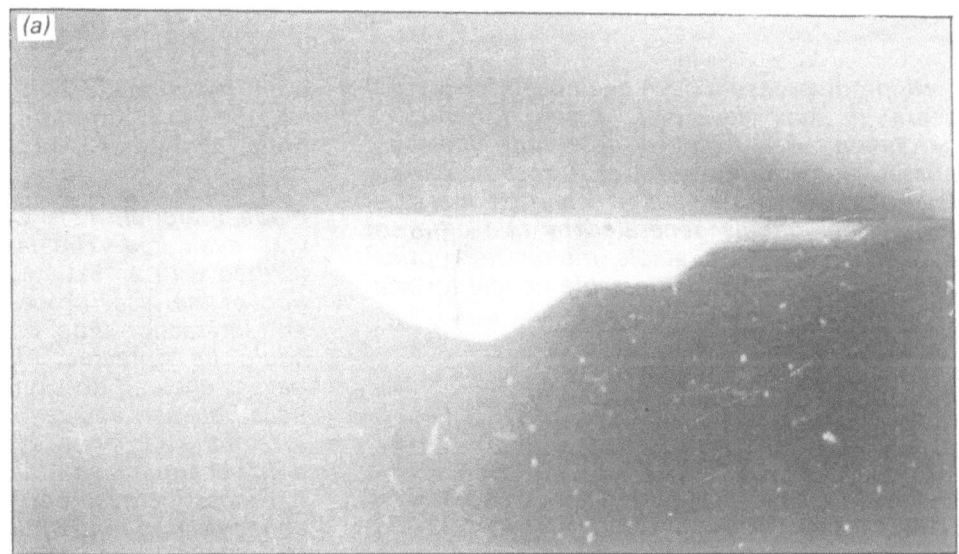

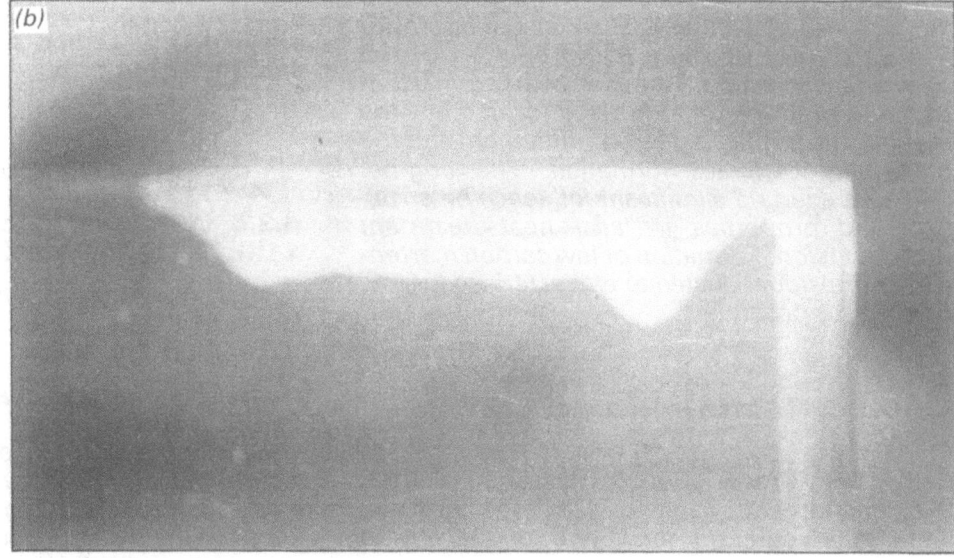

Intensity distribution of the laser beam by the stable–unstable resonator: (a) in the stable direction, (b) in the unstable direction (See 4.3.4 (15))

The 15kW AVCO laser installed at RTM institute (See 4.3.4 (17))

High-power lasers have been used for several years in heat treatment, mainly for surface hardening of steels. To determine the best operative conditions, a set of surface hardening tests has been carried out on constructional steel using laser beams generated by two different optical cavities – a stable multimode optical resonator and a stable-unstable optical resonator.

4.3.4 (16) Influence of the original structure on hardening of ShKh15 steel in treatment by CO_2 laser radiation

Safonov, A. N. et al. (Academy of Sciences of the USSR). *Met. Sci. Heat Treat.,* 27(3-4): 252-257, 1985.

Argues that in laser heat treatment with fusion the original structure has an insignificant influence on the structure and properties of the zone of laser action. Lamellar twinned martensite with high microhardness is formed but the method is complicated by embrittlement and the probability of crack formation. The original structure exerts a significant influence on structure and properties in laser heat treatment without fusion. Formation of low carbon martensite and residual austenite characterised by low hardness is possible.

4.3.4 (17) Laser hardening of a 12%-Cr steel

Roth, M. (Brown Boveri Research Center, Switzerland) and Cantello, M. (RTM Institute, Italy) In, *Proc. 2nd Int. Conf. on Lasers in*

Manufacturing, 26-28 March 1985, Birmingham, UK, (Ed. M. F. Kimmitt), pp. 119-128. IFS (Pulications) Ltd, Bedford, UK, 1985.

In this work the laser hardening of a turbine blade material, 12% Cr steel (X22CrMoV121), was evaluated. The laser treatment was performed with a 15kW AVCO laser. Due to variation of the laser processing parameters power and interaction time, different hardness profiles could be obtained. Hardening with multiple passes showed no significant drop of the hardness between adjacent passes. The microstructure of the hardened layers was characterised by light-scanning- and electron-microscopy. The hardness increase depends on the high carbon concentration in the matrix, the very fine lamellar martensite and the high dislocation density. By melting the surface with the laser, further increase in hardness is possible. In this case, alloying with carbon takes place. The carbon stems from the graphite layer which was applied in order to increase the absorptivity of the material.

4.3.4 (18) Laser hardening of 11Kh12N2V2MF steel

Lakhtin, Yu. M. et al. (Moscow Automobile and Road Institute, USSR). *Met. Sci. Heat Treat.,* 27(3-4), 247-252, 1985.

The structure, microhardness, and dimensions of the hardened zones in laser heat treatment depend on the density of the laser radiation and rate of movement of the specimen relative to the beam. In laser alloying with nitrogen they depend on the quantity of impregnating coating.

4.3.4 (19) Laser heat treatment of cast irons – optimisation of process variables. I

Mathur, A. K. and Molian, P. A. (Iowa State University, USA). *J. Eng. Mater. & Technol.*, 107(3): 200-207, 1985.

A 1.2kW CO_2 laser operating in ring and Gaussian beam modes was used to surface harden grey and ductile cast irons. The microstructure, case depth, hardness, surface integrity and distortion were studied as functions of laser power, focusing optics, beam size scan rate and thickness of the specimen. The results indicate that the ring mode is preferred to Gaussian, whereas ductile iron response to laser heat treatment is better than grey iron. The maximum case depth achieved with minimum melting was 0.38mm. (15 refs.)

4.3.4 (20) Laser heat treatment of iron-base alloys

Gnanamuthu, D. S. and Shankar, V. S. (Rockwell International Science Center, USA). In, *Proc. SPIE Applications of High-Power Lasers*, 22-23 January 1985, Los Angeles, CA, pp. 56-72. International Society of Optical Engineering, USA, 1985.

A number of techniques designed to improve the efficiency of the laser heat treatment process are presented. Methods to obtain a laser beam profile with uniform intensity involve the use of beam integrating optics with Fresnel number higher than 10. Approaches to increase beam absorptivity involve the use of infrared energy absorbing coatings or the use of linearly polarised laser beams. The numerical solution of the three-dimensional time-dependent heat conduction equation is used to calculate and experimentally validate the heat-affected and hardened zone profiles in workpieces. Core microstructure influences the hardened depth that can be obtained by laser heat treatment.

4.3.4 (21) Laser surface treatment of high-phosphorus cast iron

Zaiguang, Li. et al. (Huazhong University of Science and Technology, China). *J. Appl. Phys.*, 58(10): 3860-3864, 1985.

Samples of high-phosphorous cast iron were treated with a CW CO_2 laser to produce superficial martensitic layers by transformation hardening and superficial ledeburitic layers by surface melting. These layers were tested for resistance to cavitation erosion. Both the ledeburitic and martensitic layers were found to be less prone to cavitation erosion than the untreated surface.

4.3.4 (22) Laser transformation hardening of chromium steels: correlation between experimental results and heat flow modeling

Com-Nougué, J. and Kerrand, E. (Compagnie Générale d'Electricité, France). In, *ICALEO '84, Proc. Materials Processing Symp.*, Vol. 44, 12-15 November 1984, Boston, MA. (Ed. J. Mazumder), pp. 112-119. Laser Institute of America, Toledo, OH, USA, 1985.

The influence of the processing parameters such as the laser beam shape, the absorption coating nature, the laser power and the travel speed on the laser hardening of two 12% Cr steels has been investigated in terms of the geometry and hardness of the treated zones. The hardening trials were performed with a 5kW CO_2 laser. Besides this parametric study, the residual stresses resulting from the hardening were determined on one of the two steels and their behaviour was also compared to the hardening of a carbon steel and a low-alloy steel. Finally, a numerical solution of the heat conduction equation is presented and correlated with the experimental results.

4.3.4 (23) Properties of laser melted SG iron

Bergmann, H. W. (Universität Clausthal, W. Germany), and Young, M. (University of Birmingham, UK). In, *Proc. 2nd Int. Conf. on Lasers in Manufacturing*, 26-28 March 1985, Birmingham, UK, (Ed. M. F. Kimmitt), pp. 109-118. IFS (Publications) Ltd, Bedford, UK, 1985.

Ductile cast iron is an ideal material for laser surface melting. It is possible to produce economically a large number of components with ideal combination of a ductile tough core and a hard wear-resistant surface. The melted depth and translation rate depend on the component size, the application and the dimensional tolerance.

Macrograph of a laser melted camshaft (See 4.3.4 (23))

Cast components require deep melted layers as sufficient material must be available for the finishing operation. Finished components require only a shallow layer as finishing operation must be absent or minimal. It is shown in the paper how a crack-free, smooth (<1μm) laser melted surface can be produced. The wear resistance and fatigue of laser surface melted cast components are presented and discussed.

4.3.4 (24) Residual stresses in carbon steels after surface hardening with CO₂ laser radiation

Velikikh, V. S. et al. *Met. Sci. Heat Treat.*, 27(3-4): 258-261, 1985.

Residual compressive stresses occur in the layer of laser hardening in the X and Y directions and residual tensile stresses occur in the layer of laser tempering on the surface and in the depth. In hardening of the whole surface, both compressive and tensile stresses occur in the surface layer.

4.3.4 (25) A study of the effect of angle of beam incidence on laser transformation hardening of 4340 alloy steel

Bruck, G. J., Smith, J. E. and Nurminen, J. I. (Westinghouse Research and Development Center, USA). In, *ICALEO '84, Proc. Materials Processing Symp.*, Vol. 44, 12-15 November 1984, Boston, MA, (Ed. J. Mazumder), pp. 120-132. Laser Institute of America, Toledo, OH, USA, 1985.

A series of high-power CO₂ laser exposures of an alloy steel were conducted to study the effect of angle of beam incidence on transformation hardening. Hardening efficiencies were determined by cross-sectional measurements of the laser hardened zone. Relevance of the relationships to laser processing of intricately contoured parts is described.

4.3.4 (26) Aluminium alloy surface hardening by laser treatment

Ferraro, F., Nannetti, C. A. and Campello, M. (Istituto RTM) and Senin, A. (Olivetti, Italy). In, *Proc. 3rd Int. Conf. on Lasers in Manufacturing*, 3-5 June 1986, Paris France, (Ed. A. Quenzer), pp. 233-244. IFS (Publications) Ltd, Bedford, UK, 1986.

This work describes some results on aluminium alloys treatment by HPL to produce hard- and wear-resistant surface layers. Relatively thick layers (0.1-1mm. depending on the process parameters) have been obtained in two ways: 1) Si powder addition to form an Al-Si hypereutectit phase, due to the high thermal gradients characteristics of the laser technology. 2) Surface cladding of hard-and wear-resistant fine powders such as TiN and TiC with the aim of obtaining surface dispersion strengthening. Experiments with B₄C are now in progress.

4.3.4 (27) Laser metal surface hardening: models and measures

Kechemair, D. and Gerbet, D. (Etablissement Technique Central de l'Armement (ETCA), France). In, *Proc. 3rd Int. Conf. on Lasers in Manufacturing*, 3-5 June 1986, Paris, France, (Ed. A. Quenzer), pp. 261-269. IFS (Publications) Ltd, Bedford, UK, 1986.

Increasing diffusion of CO₂ CW high power lasers in industry, and attempts to control or robotise their applications bring out an increasing need of 'fundamental' understanding of laser-material interaction. A research programme has therefore been undertaken in this way by the 'Groupe laser' of ETCA. The paper deals with some experimental methods of dynamic data acquisition. First measurements and computation results concerning surface hardening will be qualitatively analysed.

4.3.4 (28) Laser surfacing of nickel aluminium bronze

Oakley, P. J. and Bailey, N. (Welding Institute, UK). In, *Laser Welding, Machining and Materials Processing – Proc. Int. Conf. on Applications of Lasers and Electro-Optics, ICALEO '85*, 11-14 November 1985, San Francisco, CA, (Ed. C. Albright), pp. 169-178. Laser Institute of America, Toledo, OH, USA/IFS (Publications) Ltd, Bedford, UK, 1986.

A programme of laser surface treatment of cast nickel aluminium bronze has been carried out with the ultimate objective of improving corrosion resistance. The surface treated specimens were made using a 5kW CO₂ laser. Three surface treatment techniques were used; surface remelting with a defocused beam, surface remelting with a focused beam, and surface alloying with a defocused beam using several alloying materials. The surface treated specimens were assessed by visual and metallurgical examination which allowed the microstructure and any defects to be characterised.

4.3.4 (29) Line hardening by low-power CO₂ lasers

Meijer, J. et al. (Twente University of Technology, The Netherlands). In, *Laser Welding, Machining and Materials Processing – Proc. Int. Conf. on Applications of Lasers and Electro-Optics, ICALEO '85*, 11-14 November 1985, San Francisco, CA, (Ed. C. Albright), pp. 229-238. Laser Institute of America, Toledo, OH, USA/IFS (Publications) Ltd, Bedford, UK, 1986.

When using low-power lasers only very small surfaces can be hardened. For some applications it will be enough to have only a number of hard areas in the surface to carry the mechanical load. This paper describes how hard lines are generated on C45-steel and gives results of comparative tribological tests of these surfaces. Compared with the base material a significant reduction of wear was found.

4.3.4 (30) Numerical analysis of laser surface hardening of a medium-carbon steel

Na, S. -J., Lee, S. -Y. et al. (Korea Adavanced Institute of Science and Technology). In, *Proc. 3rd Int. Conf. on Lasers in Manufacturing*, 3-5 June 1986, Paris, France, (Ed. A. Quenzer), pp. 383-392. IFS (Publications) Ltd, Bedford, UK, 1986.

The results of an FEM-simulation on laser surface hardening of a medium carbon steel are described. A two-dimensional computer program, which can be used generally for the determination of transient temperature distributions in welding and surface heat treatment, was used in the first place to investigate the effects of travel speed and beam spot diameter on the shape and size of hardened zones. For the confirmation of the accuracy of the numerical analysis a medium carbon steel of 5mm thickness was heat-treated with a 1kW CO₂ laser. A simulation scheme for the cooling time and the corresponding CCT-diagram showed that the cooling rate is high enough to consider the heated zone above 723°C as the martensitic hardening zone. With proper assumption of the absorptivity the numerical and experimental shape and size of hardened zones were in good agreement. The numerical analysis showed that the beam mode has a considerable effect on the width and depth of lenticular-shaped hardened zones and for a constant heat input per unit length the heat input rate has a great effect, if the traverse speed is low. It has also found that there exists a range of beam spot diameters, which can produce optimal hardened zones for a given beam scanning speed.

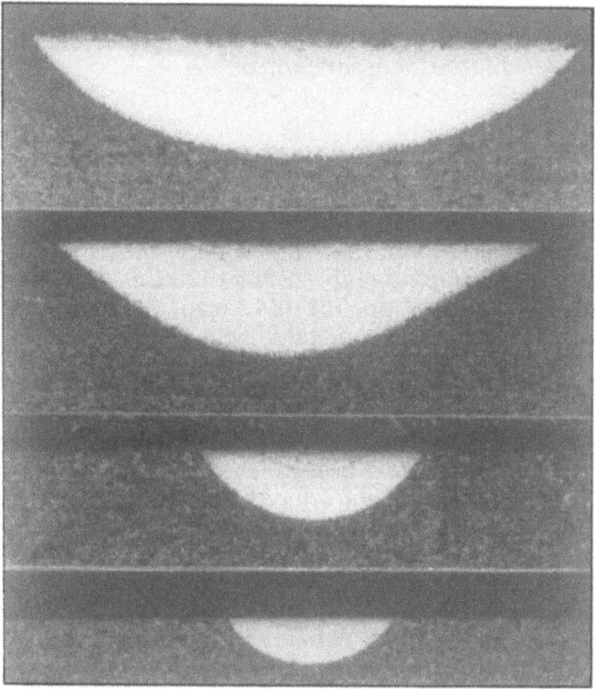

Hardened zone morphologies (See 4.3.4 (30))

4.3.5 Etching, marking and scribing

4.3.5 (1) Scribing glass with pulsed and Q-switched CO₂ laser

Saifi, M. A. and Un-Chul Paek (Western Electric, Co, USA). *Am. Ceram. Soc. Bulletin*, 52(11): 838-841, 1973.

4.3.5 (2) Lasers in the factory

Allan, R. *IEEE Spectrum*, 16(5): 42-49, 1979.

Considers industrial applications of Nd:YAG and CO₂ lasers including metal treatment, lettering, engraving, and marking fragile or small products, laser-cutting systems, trimming thin-film and thick-film resistor materials, and processing non-metals with the CO₂ laser.

4.3.5 (3) Model rough surfaces in elastohydrodynamic lubrication

Anderson, J. C., Leather, J. A. and Silva, S. D. (Metallurgical Coatings and Process Technology; San Diego, CA, 5-8 Apr. 1982, Thin Solid Films, 1 Oct. 1982, 96(1) pp. 1-8) *Thin Solid Films*, 96(1): 1-8, 1982. (Presented at Metallurgical Coatings and Process Technology Conf, 5-8 April 1982, San Diego, CA).

Model rough surfaces have been produced on steel balls by laser milling of the steel surface.

This technique was refined so that grooves of 10µm width, spaced 10µm apart, could be cut under microprocessor control.

4.3.5 (4) Production of surface textured work rolls by means of laser pulses

Bragard, A. and Crahay, J. In, *12th Biennial Congress – International Deep Drawing Research Group no. 1*, 24-28 May 1982, S. Margherita Ligure, Italy, pp. 153-164. Associacion Italiana di Metallurgia, Milan, Italy, 1982.

Describes laser texturing of rolling mill work rolls which overcomes the limitations of classical shot blasting. It provides a regular pattern of well-shaped peaks and valleys in which a certain disorder can be introduced under control. The process allows close control of the operation and is suitable for a high degree of automation.

4.3.5 (5) Target fabrication using laser and spark erosion machining

Clement, X. et al. *J. Vac. Sci. Technol.*, 20(4): 1082-1084, 1982.

Both laser and spark erosion machining were developed to produce minute parts of complex targets. A high repetition rate YAG laser at double frequency is used to etch various materials. For example, marks or patterns are often necessary on structured or advanced targets. The laser is also used to thin down plastic-coated stalks.

4.3.5 (6) Chemically enhanced laser etching

Anderson, R. B. et al. (IBM Corp, USA). *IBM Tech. Disclosure Bulletin*, 26(5): 2436, 1983.

Offers better control of surface finish and microgeometries.

4.3.5 (7) Lasers make their mark

Wildish, M. *Engineer*, 256(6618): 22-23, 1983.

Describes laser imprinting of serial numbers, brand names, bar codes, etc., on metal and other products. With the YAG lasers normally used on stationary targets the beam is focused by lenses and micro-computer-rotated mirrors and the workpiece material is vapourised at the focal point, to give dot-matrix or engraved inscriptions as desired. For broad lines and logo-type designs the mirrors are sinuousoidally rotated. Characters 3mm. high are imprinted on steel at the rate of 20/s and hard materials such as WC and stainless steel present no difficulty. The CO_2 laser is used with moving targets, such as cans on automated lines, and infra-red energy through masks or stencils gives 1500 complete marks or codes/min. In general, the smallness of the heat-affected zone obviates brittle-hardening, microcracking, and adverse thermal effects on heat-sensitive components.

4.3.5 (8) Techniques and applications of laser marking

Willis, J. B. (Laser Lines Ltd, UK). In, *Proc. 1st Int. Conf. on Lasers in Manufacturing*, 1-3 November 1983, Brighton, UK. (Ed. M. F. Kimmitt), pp. 53-62. IFS (Publications) Ltd, Bedford, UK, 1983.

Non-metallic materials can be marked with CO_2 lasers with either a pulsed output producing a complete mark in a single shot or continuous output with raster scanning. Typical uses of pulsed CO_2 lasers include "BEST BEFORE" marks on food packaging and type numbers on plastic packages. Scanning CO_2 lasers are used most often for rubber and wood. Practically all types of material can be marked with Nd:YAG lasers which can be pulsed to produce dot matrix patterns or continuously pumped to give high writing speeds using moving optics with either X-Y translation tables or galvanometer-driven mirrors. Applications of YAG lasers include part marking in aerospace, automotive, medical and nuclear industries amongst many others. Laser marking does not replace conventional marking but supplements it where speed, versatility and good presentation are required. The particular technique used should be carefully chosen to suit the application.

4.3.5 (9) Mill roll engraving by laser for better product consistency

Steel Times Int., 8(2): 55, 1984.

Describes precisely controlled lasertex surface textures impressed on metal sheet by laser-engraved rolls. Reproducibility and other advantages as compared with shot blasting are listed. The engraving is carried out by an automatically interrupted laser beam traversing a rotating roll, to give impact points forming craters along a helicoidal path. The depth and spacing of the craters are functions of the beam intensity and the roll-rotation rate, thus roll-surface relief is perfectly controlled.

4.3.5 (10) Drilling, marking and other applications for industrial Nd:YAG lasers

Tiffany, W. B. (Coherent General Inc. USA). In, *Proc. SPIE Applications of High-Power Lasers*, Vol. 527, 22-23 January 1985, Los Angeles, CA, pp. 28-36. International Society of Optical Engineering, USA). 1985.

Describes coherent general's new EVERPULSE M34 laser machine as an example of significantly improved beam quality in a high power Nd:YAG drilling laser. Drilling requirements are related to laser variables including pulse energy and duration, number of pulses, focusing parameters, and output beam quality. Performance of Nd:YAG, Nd:glass, and face pumped TIR slab geometry lasers are compared.

4.3.5 (11) Excimer lasers in photolithography

Gower, M. C. (SERC Rutherford Appleton Laboratory, UK). *In, Proc. 2nd Int. Conf. on Lasers in Manufacturing*, 26-28 March 1985, Birmingham, UK, (Ed. M. F. Kimmitt), pp. 67-72. IFS (Publications) Ltd, Bedford, UK, 1985.

Because of their high pulsed powers in the ultraviolet spectral region, excimer lasers are of interest as lamp sources in photolithographic mask aligners. Their short wavelengths enable higher resolution and packing densities to be achieved on the silicon chip, while their high power should allow large wafer areas to be processed with a rapid throughput of devices.

4.3.5 (12) Lasers in practice. IV

Luft, A. *Elektron. Prax.*, 20(10): 143-144, 1985. (In German) (For pt. III see ibid., 20(9): 113, 1985).

Describes some uses of the laser in fabrication processes, including the use of a focused laser beam to evaporate material along a predetermined line for scribing. Laser scribing of both silicon wafers and aluminium oxide are discussed. Concludes with a discussion of laser beam trimming and shaping of materials.

4.3.5 (13) Polymer film cutting and ablative etching using a 1kHz XeC1 laser

Bishop, G. J. and Dyer, P. E. (Hull University, UK). *Appl. Phys Lett.*, 47(11): 1229-1331, 1985.

Significant increases in cutting efficiency and etch rate occur at high pulse rates due to cumulative heating effects. Cut rates of 130cm s-1 are obtained at 900Hz for a 12nm thick polyethylene teraphthalate film. (10 refs.)

4.3.5 (14) Programmable laser character generation

Greenwood, D. I. et al. (Isomet Laser Systems Ltd, UK). In, *Proc. 2nd Int. Conf. on Lasers in Manufacturing*, 26-28 March 1985, Birmingham, UK, (Ed. M. F. Kimmitt), pp. 279-284. IFS (Publications) Ltd, Bedford, UK, 1985.

Acousto-optic techniques for the generation of spot matrix alpha-numerics on continuously moving target material are described. The systems utilise a high p.r.f. Nd:YAG laser to mark the characters point by point.

4.3.5 (15) Present state of development of the 'Lasertex' process

Crahay, J. et al. (Centre de Recherches Métallurgiques (CRM) Belgium). In, *Proc. 3rd Int. Conf. on Lasers in Manufacturing*, 3-5 June 1986, Paris, France, (Ed. A. Quenzer), pp. 245-260. IFS (Publications) Ltd, Bedford, UK, 1986.

It was in 1982 that for the first time rolls fitted with a roughness textured by laser beam were used on industrial rolling mills. Pilot productions of coils were achieved at that time by Cockerill-Sambre and by Sidmar. The rolls were textured in a pilot plant situated in a workshop of OSB and operated by a CRM team. Since then Sidmar and Cockerill have acquired their own texturing installations. The basic design of these two installations as well as the construction of the engraving heads had been committed to CRM. CRM also brought in its scientific and technical support when particular problems arose. Nowadays it is possible to draw some information from the industrial experience gained with this new engraving process, called Lasertex.

4.3.5 (16) Sub-contracting for laser cutting, welding and engraving

Nicolas, S. (Laser Industrie, France) In, *Proc. 3rd Int. Conf. on Lasers in Manufacturing*, 3-5 June 1986, Paris, France, (Ed. A. Quenzer), pp. 355-358. IFS (Publications) Ltd, Bedford, UK, 1986.

Presents the advantages of sub-contracting in the laser machining domain. Readers are reminded of the advantages of the two complementary laser technologies (CO_2 laser and YAG laser) utilised by Laser Industrie for cutting, boring and welding. The advantages of a specialised sub-contracting and a real catalyser of applications in service to industry are also presented.

4.4 METALLURGICAL ANALYSIS OF LASER PROCESSES

4.4.1 General

4.4.1 (1) Properties and structures of laser welds of high-strength alloys

Metzbower, E. A. and Moon, D. W. In, *Proc. Weldments: Physical Metallurgy and Failure Phenomena Conf.*, 27-30 August 1978, Lake George, NY, pp. 245-256. General Electric Co., Schenectady, NY, 1979.

A series of laser welds of 12 mm. thick plates of high-strength steel, Al and Ti were fabricated and tested. Mechanical properties: yield and ultimate strengths, elongation, fracture resistances, and reduction-in-area were measured and calculated. Fractographic observations were made on both tensile and DT test specimens. Metallographic examinations were made in the base plate, fusion and heat-affected zones. Correlations between structures and properties are made.

4.4.1 (2) Evaluation of residual stresses in welded joints of thin plate by a holographic method

Rassokha, A. A. and Talalaev, N. N. *Ind. Lab.*, 48(11): 1143-1146, 1982. Also *Strength of materials*, 15(1), 1983.

A description is given of a method of holographic investigation of residual stresses in welded joints of plates produced by argon-arc and laser welding. It is concluded that the holographic method can be used in the investigation of residual stresses in welded joints of thin plates produced by laser welding. Simple calculation equations are developed using holographic methods to determine the residual stresses in plates with a two-dimensional residual stress system.

4.4.1 (3) Structural defects in laser and electron beam annealed silicon

Narayan, J. (Oak Ridge National Laboratory, USA). Rept. no. CONF-791112-62, 14pp., 1979. Presented at *Symp. on the Scientific Basis for Nuclear Waste Management*, 26 November 1979, Boston, MA.

Microstructural modifications obtained as a function of laser parameters are examined and it is shown that laser beam pulses can be used to remove displacement damage, dislocation loops and precipitates. Annealing of defects underneath the oxide layers in silicon is possible within a narrow energy band.

4.4.1 (4) Laser welding of structural alloys

Fraser, F. W., Metzbower, E. A. and Moon, D. W. In, *Conf. on Welding Technology for Energy Applications*, 16-19 May 1982, Gatlinburg, TN, pp. 313-329. Oak Ridge National Laboratory, Oak Ridge, TN, USA, 1982.

Laser beam weldments of structural alloys have been fabricated in a single pass, both as autogeneous butt welds and narrow gap Vee welds with filler metal additions. Aluminium alloys, titanium alloys and steels were welded at power levels from 10 to 14kW and speeds from 13 to 19mm/s. For all of the alloys except Al the mechanical properties corresponded to the base plate properties. In the case of Al the weldments fractured in the fusion zone. The requirements of a bend test were satisfied by the weldments. Hardness tests were made across the base plate, heat affected and fusion zones, and microstructures were determined in the three zones and correlated with the mechanical properties and hardnesses. Fusion zone purification was observed in certain alloys (17 refs.)

4.4.1 (5) Special features of the solidification of laser welds

Gavrilyuk, V. S.; et al. *Automat. Weld.*, 35(6): 28-30, 1983.

4.4.2 Metallurgy of laser welding and processing of ferrous metals and alloys

4.4.2 (1) A metallurgical characterisation and assessment of SMA, GMA, EB, and LB welds of HY-130 steel

Stoop, J. and Metzbower, E. A. (Naval Research Laboratory, USA). Rept. No. NRL-8157, 56pp; 1977. Naval Research Laboratory, Washington, DC, USA.

HY-130 weldments of 6.35 and 12.7mm thickness were fabricated by shielded metal arc, gas metal arc, electron beam, and laser beam processes. Hardness explorations and metallographic examinations were made of the weld metal and heat-affected zone of the weldments. EB and LB weldments showed perceptibly higher weld metal hardnesses and steeper hardness gradients in the HAZ than corresponding SMA and GMA weldments. EB and LB welds comprised mostly martensite and a small percentage of bainite. Strain hardening exponent values for

SMA, GMA, and LB welds were appreciably higher than the values obtained for EB welds and the base metal. The weld joint specimens exhibited mostly plane stress (slant) fracture. LB specimens disclosed a fracture toughness not only higher than that of the other weld joints but also comparable to the fracture toughness of the base metal. LB weld joint fracture energy values were found to be 94.5 to 97.7 percent of those of the base metal. There was some evidence of cold shuts and hydrogen embrittlement in the LB weldments.

4.4.2 (2) Effect of laser welding on weld metal structure

Fedorov, V. G. et al. (Moscow Higher Technical School, USSR). *Izv Vyssh Uchebn Zaved Mashinostr*, (2): 122-125, 1979. (In Russian)

18Cr2Ni3MoCu steel weld metal structure obtained by laser welding at welding speeds of 40 and 100m/h is discussed. It is shown that laser welding results in a fine grained weld metal structure, and a narrower zone of hardening and tempering, especially for high welding velocities. Weld metal nonuniformity, is lower in the case of laser welding than in argon arc welding.

4.4.2 (3) Study of residual stresses and distortion in structural weldments in high strength steels

Papazoglou, V. J. and Masubuchi, K. (Institute of Technology, USA). Rept. No. 82558, 99pp; November 1979.

Presents experimental results for welding HY-130 steel. The geometries include 1in. thick plates and 0.75in. cylindrical shells. Welding was performed using the gas metal arc, electron beam and laser processes. The experimental results are compared with predictions made by computer programs. Through thickness distributions of residual-stresses are presented.

4.4.2 (4) Comparison of technological strength of welds prepared by beam and arc welding methods

Andriyakhin, V. M. *Avtom. Svarka.*, 33(10): 9-12, 1980. (In Russian)

The effect of high heat energy concentration tendency to form hot and cold cracks was studied on laser welded steels. Weld resistance to hot crack decreased with increasing welding rate. The beam welding methods, in comparison to Ar arc welding, gave higher resistance to cold cracking, in spite of the martensite formed in the weld and in the heat affected zone. (9 refs.)

4.4.2 (5) Evaluation of laser welding techniques for hydrogen transmission

Mucci, J. (Pratt and Whitney Aircraft Group, USA). Final Rept. No. COO-4355-11, 85pp; May 1980. See also Rept. No. 4355-7/8/9/10, 1978-1979. Pratt and Whitney Aircratt Group, West Palm Beach, FL, USA.

This programme was established to determine the feasibility of laser beam welding as a fabrication method for large-scale, economic pipeline transmission of fuel for a hydrogen energy system. Considers the effect of conventional weld processes and laser beam welding on the mechanical properties of two classes of steels in an air and high pressure gaseous hydrogen environment. Evaluation of the tensile, low-cycle fatigue and fracture toughness properties and metallurgical analyses provide the basis for concluding that laser beam welding of AISI 304L stainless steel and ASTM A106B carbon steel can produce weldments of comparable quality to those produced by gas-tungsten arc and electron beam welding.

4.4.2 (6) Analytical electron microscopy evaluation of laser welded 308 stainless steel

Vitek, J. M. and David, S. A. (Oak Ridge National Laboratory, USA). In, *Scanning Electron Microscopy*, pt 1, pp.l 169-175. Scanning Electron Microscopy Inc, Chicago, IL, USA, 1981. Also ORNL Conf Rept. No. CONF-810410-3, 20pp; 1981.

Due to the small amounts and fine scale of the minor phases found in the laser welded material, AEM is unique in its ability to determine compositions accurately. The laser welded microstructures were different from those produced by more conventional techniques in that the amount of delta ferrite found in an austenite matrix was much lower. Additionally, a uniform distribution of amorphous Mn-Si rich precipitates was found. Compositions were determined as a function of both welding speed and post-weld aging.

4.4.2 (7) Evaluation of possibility to increase the service reliability of steel 08Kh15N5D2T laser welded joints

Weld. Prod., 28(11): 15-17, 1981.

Comparative investigation of structure, strength and corrosion resistance was done on the welds of martensitic stainless steel 08Kh15N5D2T carried out by laser welding. The structure of the weld metal and of the heat-affected zone was investigated and the hardness and amounts of retained austenite were also measured. Laser welded joints were obtained without a tendency to stress corrosion cracking and had fine microstructure, narrow heat-affected zone free of carbide grid.

4.4.2 (8) Microstructural analysis of austenitic stainless steel laser welds

Vitek, J. M. and David, S. A. (Oak Ridge National Laboratory, USA). Rept. No. CONF-811184-4, 19pp; 1981. Oak Ridge National Laboratory, TN, USA. Presented at *Conf. on Trends in Welding Research in the US*, 16 November 1981, New Orleans, LA.

Analysis of laser welded type 308 stainless steel shows that the high cooling rates encountered in the process have noticeably altered the structures. Results show promise in solving the common problems of hot cracking and elevated temperature embrittlement in austenitic stainless steels welds. The solidification mode is changed from one of primary ferrite formation to primary austenite. This change in solidification mode results in a large reduction in the ferrite level to less than 1%. Although the welds are nearly fully austenitic, no evidence of hot cracking was found. (See also other later papers by these authors.)

4.4.2 (9) Solidification behaviour and microstructural analysis of austenitic stainless steel laser welds

David, S. A. and Vitek, J. M. (Oak Ridge National Laboratory, USA). In, Lasers in Metallurgy, Chicago, IL, 22-26 Feb. 1981, Metallurgical Society/AIME, Warrendale, PA, 1981. In, *Proc. Lasers in Metallurgy Conf.*, 22-26 February 1981, Chicago, IL, pp. 247-254. Metallurgical Society/AIME, Warrendale, PA, USA, 1981. Also available as Rept. no. CONF-810203-13, 8pp., 1981.

Solidification behaviour of 308 austenitc stainless steel laser welds has been investigated with a high-power laser system. The welds were made at speeds ranging from 13 to 60mm/s. The welds showed a wide variety of microstructural features. The ferrite content in the 13mm/s weld varied from < 1% at the root of the weld to approximately 10% at the crown. However, the welds made at 25 and 60mm/s. contained an austenitic structure with < 1% ferrite throughout the weld. The austenitic stainless steel welds were free of any cracking; the results are explained in terms of the rapid solidification conditions during laser welding. (8 refs.)

4.4.2 (10) Transformation and tempering behaviour of 12Cr-1Mo-0.3V Martensitic stainless steel weldments

Lippold, J. C. *J. Nucl. Mater.*, 104(1-3): 1127-1131, 1981. Presented at *Conf. on Fusion Reactors*, Part B, 9-12 August 1981, Seattle, Washington.

Autogenous, bead-on-plate laser welds in a 12Cr-1Mo-0.3V (HT9) martensitic stainless steel were evaluated. The as-welded fusion zone microstructures consisted of a mixture of untempered martensite and metastable Delta ferrite and exhibited a hardness in the range of R_c 48-55. The tempering behaviour of the laser welds was similar. The tempering response was relatively sluggish at temperatures < 600 deg C. Tempering for 1h at 800°C reduced the hardness of both the fusion zone and HAZ to the level of the quenched and tempered base metal.

4.4.2 (11) Effect of the type of welding on the corrosion resistance of welded structures made of 03Kh11N10M2T and 08Kh15N5D2T steels

Svar. Proizvod., (11): 26-27, 1982. (In Russian)

Corrosion resistance was studied in martensitic 03Kh11N10M2T (steel A) and 08Kh15N5D2T (steel B) structures welded by laser, electron beam and argon arc. Mechanical and corrosion properties are tabulated. Various welded structures made of both steel A and B had higher corrosion resistance when laser welded in moisture chambers than when electron beam or arc welded.

4.4.2 (12) Laser welding of structural and stainless steels

Hill, M. and Megaw, J. H. C. In, *ICALEO '82, Proc. Materials Processing Symp.*, Vol. 31, 20-23 September 1982, Boston, MA, pp. 108-115. Laser Institute of America, Toledo, OH, USA, 1982.

Discusses metallurgical and process considerations which arise in the welding of structural and stainless steels. In structural steel the use of filler to control problems of fit-up and metallurgy is considered and some thick-section autogeneous welds are considered, particularly their impact properties. In the stainless steels, the emphasis is placed on fabrication of specific component shapes and aspects concerned with closure of circular welds, an application in which the low, reproducible shrinkage and distortion have been exploited. (10 refs.)

4.4.2 (13) Microstructural analysis of austenitic stainless steel welds and their modification during laser welding

David, S. A., Vitek, J. M. and Smith, W. H. (Oak Ridge National Laboratory, USA). Rept. No. ORNL/TM-8276, 30pp., May 1982. Oak Ridge National Laboratory, TN, USA, 1982.

Four distinct ferrite morphologies were iden-tified in 308 stainless steel weld metal: vermicu-lar, lacy, acicular, and globular. The first three ferrite types are related to transformations dur-ing and after solidification, and the fourth is related to the shape instability of the residual ferrite. Solidification behaviour of the steel welds at high cooling rates was investigated with a high-power laser system. Welds made at 13mm/s.contained microstructures ranging from a fully austenitic structure at the root of the weld to a ferrite plus austenitic duplex structure at the crown. However, the welds made at 25 and 63mm/s. contained a nearly fully austenitic structure with less than 1% ferrite, and third-phase particles rich in manganese and silicon were also found. The fully austenitic stainless steel welds were free of any cracking. The results are explained in terms of excessive undercool-ing at the tip of the ferrite cells at rapid solidification rates.

4.4.2 (14) A study of the dependence of microsegregation on critical solidification parameters in rapidly-quenched structures

Snow, D. B., Greenwald, L. E. and Breinan, E. M. (United Technologies Research Center, USA). Rept. No. UTRC/R82-914763-6, 55pp., January, 1982. United Technologies Research Center, East Hartford, CT, USA.

Grade AH32 structural hull steel clad with varying thicknesses of 317L stainless steel were bead-on-plate welded with a continuous CO_2 laser operated at power settings of from 2 to 9kW and scanned at 5 to 76cm/sec. The resulting fusion zone shapes and microstructures were examined to determine the variation in welding efficiency and fusion zone segregation as a function of beam power input and cladding thickness. Due to the similarity in thermal con-ductivity and reflectivity of the cladding and substrate, the degree of the dilution of chro-mium and nickel in the fusion zone depended only on the relative volumes of cladding and substrate mixed in it, and not the absolute values of cladding thickness or fusion zone depth. Welding efficiency increased with power and beam traverse speed at constant fusion zone depth.

4.4.2 (15) Corrosion fatigue crack growth rates of weldments in Q1N steel

Jones, B. F. and Galsworthy, J. C. (Admiralty Marine Technology Establishment, UK). Rept. No. AMTE(M)-R83003, 33pp., June 1983. Admir-alty Marine Technology Establishment, Poole, UK.

Low frequency fatigue crack growth rates in air and sea water are reported for manual metal arc, synergic metal inert gas, electron beam and laser beam welds in Q1N steel. The stress corrosion crack growth susceptibility in sea water of the weldments was also studied and both the electron beam and laser beam welds exhibited susceptibility whilst the other welds and parent steel did not. All welds except the 'as-welded' laser beam welds showed similar fatigue crack growth rates to the parent plate. When tested in sea water the laser beam weld was found to exhibit exceptionally fast crack growth rates. The factors influencing fatigue crack growth in weldments are discussed.

4.4.2 (16) The effect of trace elements on GTA weld penetration

Burgardt, P., Heiple, C. R. and Roper, J. R. In, *Proc. Modelling of Casting and Welding Proces-ses II*, 31 July - 5 August 1983, Henniker, NH, pp. 193-205. Metallurgical Society/AIME, Warren-dale, PA, USA, 1984.

Describes a surface tension driven fluid flow model for the mechanism by which trace ele-ments affect GTA weld pool shape. It is prop-osed that fluid flow in the weld pool is generally the major factor determining weld pool shape. Trace elements alter fluid flow patterns by changing surface tension gradients on the weld pool surface. The effects of sulphur, oxygen, Se, Te, Al, and Ce on weld shape in 21-6-9 stainless steel are presented and compared with the predictions of the model. The change in GTA and defocused-laser weld shape from low-level Se doping is similar. (44 refs.)

4.4.2 (17) Materials-science aspects of laser cutting

Barton, G., Koschlig, M. and Bergmann, H. W. (Technische University Clausthal, W. Germany). *Z Werkstofftech*, 14(8): 257-263, 1983. (In Ger-man)

Discusses the techniques of laser cutting and the microstructural changes it causes in the material. It is shown which cutting speeds can be used for different sheet thicknesses and how this changes the surface roughness and hardness. The influence of laser power and cutting gas pressure (oxygen) is demonstrated. Studies were made of low-alloy carbon steels, and of high-alloy austenitic, ferritic and martensitic steels.

4.4.2 (18) Mechanical properties and acoustic emission in laser welded HSLA steel

Dionoro, G. and Teti, R. *J. Acoust. Emiss.*, 2(4): 281-287, 1983.

Uses a 15kW continuous wave CO_2 laser. The AE response was correlated with working parameters and mechanical characteristics of the welded joint. On the basis of the correlations between mechanical properties and laser parameters, AE patterns from the various types of welded samples are shown to be dependent on weld quality. (7 refs.)

4.4.2 (19) Microstructural modification of austenitic stainless steels by rapid solidification

Dasgupta, A., David, S. A. and Vitek, J. M. *Metall. Trans. A*, 14A(9): 1833-1841, 1983.

308, 310 and 312 steels were evaluated after rapid solidification. These three steels are commonly-used weld filler metals. Two methods of rapid solidification were investigated: autogenous laser welding and arc-hammer splat quenching. The structure of 310 stainless steel was found to be 100% austenite and did not vary over the range of conditions studied. The structures of 308 and 312 steels were sensitive to the cooling rates and solidification conditions. With the highest cooling rates, the 308 structure was fully austenitic while the 312 structure was fully ferritic. At lower cooling rates, the structures were duplex ferrite plus austenite. (24 refs.)

4.4.2 (20) Determination of weld pool temperature during laser welding of AISI 202

DebRoy, T. and Khan, P. A. A. (Pennsylvania State University, USA). In, *Proc. Laser Processing of Materials*, 26 February- 1 March 1984, Los Angeles, CA, pp. 71-81. Metallurgical Society/AIME, Warrendale, PA, USA, 1985. Also *Metall. Trans. B*, 15B(4): 641-644, 1984.

Demonstrates that the relative rates of vapourisation of any two elements from the molten pool can serve as an indicator of weld pool temperature. Alloying element vapourisation rates, composition of the vapourised material and the weld pool temperature were determined for various experimental conditions during laser welding of AISI 202 stainless steel. The vapourised material consisted primarily of Fe, Mn and Cr. The composition of the solidified region calculated from the measured values of vapourisation rate, plasma composition and the volume of the solidified region was in good agreement with the weld composition determined by electron probe microanalysis. (9 refs.)

4.4.2. (21) CO_2 laser welding of deep drawing steel sheet and microalloyed steel plate

Dawes, C. J. and Watson, M. N. (Welding Institute, UK). In, *ICALEO '83, Proc. Materials Processing Symp.*, Vol. 38, 14-17 November 1983, Los Angeles, CA, (Ed. E. A. Metzbower), pp. 73-79. Laser Institute of America Toledo, OH, USA, 1984.

Laser welding trials have been conducted on 2 to 4mm. (0.080 to 0.160in.) low-carbon deep drawing steel in both butt and T joint configurations, and on butt joints in 12.5mm. (0.490in.) microalloyed steel. A 5kW fast axial flow CO_2 laser was used. The incidence of defects and the weld structures and properties are described. (5 refs.)

4.4.2 (22) Fundamental research on laser welding on structural steel

Arata, Y. and Oda, T. *Jpn. Weld. Res. Inst.*, 13(2): 227-233, 1984. Also *J. High Temp. Soc. Jpn.*, 10(3): 118-125, 1984. (In Japanese)

The laser weldability of structural steel was investigated along with the various fundamental characteristics of laser welding. The mechanical characteristics of the weld are satisfactory.

4.4.2 (23) HSLA steel laser beam weldments

Denney, P. E. and Metzbower, E. A. (Naval Research Laboratory, USA). In, *ICALEO '83, Proc. Materials Processing Symp.*, Vol. 38, 14-17 November 1983, Los Angeles, CA, (Ed. E. A. Metzbower), pp. 80-86. Laser Institute of America, Toledo, OH, USA, 1984.

A 15kW laser was used to produce autogenous welds in 12mm. (0.5in) thick plates of three types of high strength low-alloy (HSLA) steels (ASTM A633, A710/736, and A737). Mechanical properties (YS, UTS, Elong., RA and fracture toughness) were determined for the base plate and weldments. Microhardness traverses were also made from the base plate through the heat affected zone (HAZ) and fusion zone. CVN values of autogenous welded ASTM A710/736 were compared to those of GMA welded and laser welded/Inconel alloyed fusion zone weldments. Also examined were the CVN values of the HAZ. A discussion on laser welding and how its rapid cooling affects microstructure and mechanical properties in HSLA steels will be presented. (12 refs.)

4.4.2 (24) Microstructure of ASTM A-36 steel laser beam weldments

Strychor, R., Moon, D. W. and Metzbower, E. A. (Carnegie-Mellon University, USA). *J. Met.*, 36(5): 59-61, 1984.

Laser welding produces a bainitic fusion zone and a refined ferrite/pearlite heat-affected zone, resulting in improved mechanical strength and toughness compared to the base plate. The stress-relief heat treatment of 1175 degree f/1h breaks up and spherodises the carbide films at the bainite lath boundaries. This increases the impact energy the material can absorb down to low temperatures. Laser welding can be used to produce weldments with good mechanical properties where high cooling rates refine the microstructure without the embrittling martensite found in high-carbon steels. The toughness can be increased further through the use of a stress-relief anneal which coarsens and spherodises the bainitic carbides.

4.4.2 (25) Microstructure of rapidly quenched type 308 stainless steel weld filler metal and its implications on rapid solidification processes

David, S. A. and Vitek, J. M. (Oak Ridge National Laboratory, USA). Rept. no. CONF-8409107-6, 5pp., September 1984. Oak Ridge National Laboratory, TN, USA. Presented at Int. Conf. on Rapidly-Quenched Metals, 3 September 1984, Wurzburg, W. Germany.

A large degree of variation was found from a fully austenitic structure to the duplex ferrite plus austenite structure commonly found in welded material. The lack of large-scale segregation indicates the structural variations are due to cooling rate changes rather than compositional fluctuations.

4.4.2 (26) Results of underwater welding with high-power CO_2 lasers

Sepold, G. and Teske, K. (BIAS Bremer Institute, W. Germany). In, *ICALEO '83, Proc. Materials Processing Symp.*, Vol. 38, 14-17 November 1983, Los Angeles, CA, (Ed. E. A. Metzbower), pp. 87-89. Laser Institute of America, Toledo OH, USA, 1984.

Laser welding and its applicability for underwater wet welding are studied. Questions arise as to whether, due to the rapid heating and quenching rates by laser welding, harmful influences of the surrounding medium water on the welding result can be avoided. Hydrogen embrittlement of the joint should be reduced to a minimum. Preliminary experiments on the laser beam welding underwater had been carried out *and the welded seams tested as to H acceptance, embrittlement and formation of the seam. Low-carbon St52-3 steel plate 6mm. thick was used in the experiments.*

4.4.2 (27) Special features of the metallurgy of laser welding

Surkov, A. V., Kovalev, V. V. and Novozhilov, N. M. *Weld. Prod.*, 31(5): 15-16, 1984.

Presents results of investigations into the transfer of carbon silicon, manganese, and titanium from the parent metal and also of nitrogen and oxygen from the gas phase into the metal of pearlitic and austenitc laser-welded joints and compares them with those obtained with other types of fusion welding. The extent of transfer of nitrogen from the gas phase into the weld metal in laser welding of steels is lower than that in arc welding in shielding gases.

4.4.2 (28) Structure and mechanical properties of dissimilar joints made by laser welding

Fedorov, V. G. et al. *Autom. Weld.*, (9): 44-46, 1984.

Studies composite joints made of 35 steel and 12Kh2N4A alloy steel by laser, electron beam and Ar-arc welding. Laser welding resulted in a fine-grained structure in the latter steel, with the least amount of quenched and tempered portions in the heat affected zone of this steel. The structure control achieved by laser welding resulted in the highest impact resistance and cold cracking resistance in the 12Kh2N4A steel. However, for the 35 steel, the high cooling rates typical of laser welding resulted in the formation of large needles of martensite with high microhardness and low ductility, giving poor impact resistance and some cold crack formation.

4.4.2 (29) Experimental and theoretical studies on transport processes in laser welding

DebRoy, T. (Pennsylvania State University, USA). Rept. No. DOE/ER/45158-TI, 3pp., March 1985.

Progress has been made in areas related to fluid flow, heat transfer and alloying element loss during laser welding of stainless steels. Analysis at weld pool temperature, alloying element vapourisation rate, and solute loss; absorption of CO_2 laser by stainless steels; numerical calculation of fluid flow in laser melted pools; and microstructure of the welded material is reported.

4.4.2 (30) Solidification behaviour of laser welded stainless steel

Molain, P. A. (Iowa State University, USA). *J. Mater. Sci. Lett.*, 4(3): 281-283, 1985.

Laser welding of AISI 304 austenitic stainless steel was performed to determine its effect on the mode of solidification, morphology of solidification structure, hot cracking tendency, microsegregation and microstructure. It was found that the steel exhibited cellular-dendritic structures, solidified with delta -phase as the primary product of solidification, and showed complete absence of hot cracks, delta -ferrite and microsegregation.

4.4.2 (31) Solidification microstructure of laser welded stainless steels

Katayama, S. and Matsunawa, A. (Osaka University, Japan). In, *ICALEO '84, Proc. Materials Processing Symp.*, Vol. 44, 12-15 November 1984, Boston, MA, (Ed. J. Mazumder), pp. 60-67. Laser Institute of America, Toledo, OH, USA, 1985.

The solidification microstructures of pulsed Nd:YAG laser or CW CO_2 laser welds were studied in commercial SUS 310S and 304 austenitic stainless steels and in several other austenitic and duplex stainless steels. In the pulsed laser welding of Type 310S, neither grain growth nor recrystallisation grain-refining was observed in the heat-affected zones (HAZ), and planar and cellular growth occurred epitaxially in weld metals from adjacent, incompletely-melted grains in the HAZ. According to the data relating a primary dendrite arm spacing to the cooling rate in Type 310S weld metals, the average local cooling rates within the single laser shot welds were extrapolated to be so rapid as to range from 5×10^4 to $5 \times 10^{6}°C/s$ depending mainly on the pulse energy. In the case of Type 304, the pulsed laser weld metals were almost fully austenitic in contrast to GTA or CO_2 laser weld metals containing about 5% residual delta-ferrite. It was significantly confirmed that almost fully austenitic structure of Type 304 could be produced with extremely high cooling rates obtained by CW CO_2 laser welding process with low heat inputs at high traverse speeds. From the microstructural observation of several materials, it was further revealed that the solidification microstructures of pulsed laser welds were not consistent with the prediction from the Schaeffler diagram. (24 refs.)

4.4.2 (32) The effects of sulfur and nickel on the mechanical properties of HY-steel laser weldments

Moon, D. W. and Metzbower, E. A. (Naval Research Laboratory, USA), and Phillips, R. H. (Materials Research Laboratories, Australia). In, *Laser Welding, Machining and Materials Processing – Proc. Int. Conf. on Applications of Lasers and Electro-Optics, ICALEO '85*, 11-14 November 1985, San Francisco, Ca, (Ed. C. Albright), pp. 3-10. Laser Institute of America, Toledo, OH, USA/IFS (Publications) Ltd, Bedford, UK, 1986.

HY-80 and HY-100 steels of normal (0.012 wt.%) and low (0.002 wt. %) Sulphur content were laser welded with and without nickel insert in 12mm. (0.5in) thick plates.

Charpy V-notch tests on all weldments were conducted over the temperature range (-90°F to 68°F). In general, welds with low sulphur levels gave significantly higher toughness compared to welds with normal sulphur levels both in HY-80 and HY-100. An alteration in this trend, however, was noted in that low S HY-100 did not show a toughness improvement at temperatures below -51°C (-60°F). A strong synergistic effect between low S and Ni inoculated welds was well reflected in remarkable improvement in low temperature toughness for both HY-80 and HY-100 welds. The mechanical behaviour of all welds has been correlated with corresponding welded structures and fractography of the fracture surfaces.

4.4.2 (33) The role of heat input in deep penetration laser welding

Carlson, K. W. (Westinghouse Corporate Laser Center, USA). In, *Laser Welding, Machining and Materials Processing – Proc. Int. Conf. on Applications of Lasers and Electro-Optics, ICALEO '85*, 11-14 November 1985, San Francisco, CA, (Ed. C. Albright), pp. 49-57. Laser Institute of America, Toledo, OH, USA/IFS (Publications) Ltd, Bedford, UK, 1986.

Full-penetration welds were fabricated in 12.7mm (1/2in.) type 304 stainless steel using a number of different laser welding parameters to determine the effect of heat input on the resultant welds. The weld characteristics considered were bead profile, depth of penetration, integrity, and microstructure. Finally, the implications and limitations of the results are discussed. (15 refs.)

4.4.2 (34) Solidification behaviour and microstructural characteristics of pulsed and continuous laser welded stainless steels

Katayama, S. and Matsunawa, A. (Osaka University, Japan). In, *Laser Welding, Machining and Materials Processing – Proc. Int. Conf. on Applications of Lasers and Electro-Optics, ICALEO '85*, 11-14 November 1985, San Francisco, CA, (Ed. C. Albright), pp. 19-25. Laser Institute of America, Toledo, OH, USA/IFS (Publications) Ltd, Bedford, UK, 1986.

This study was undertaken by using about 25 different Fe-Cr-Ni ternary stainless steels to obtain a better understanding of the complicated microstructure observed in pulsed YAG laser and CW CO_2 laser weld metals. On the basis of the interpretation of Fe-Cr-Ni ternary diagram and the microstructural observation of GTA weld metals quenched rapidly during welding, stainless steel weld metals were classified into five different types of solidification processes. In this way, the proposed solidification modes were useful in interpreting the ferrite-austenite morphologies and solidification process in stainless steel weld metals subjected to a wide variety of cooling rates. (15 refs.)

4.4.3 Metallurgy of laser welding and processing of non-ferrous metals and alloys

4.4.3 (1) Pulsed laser welding of molybdenum

Jellison, J. L. (Sandia Laboratories, USA). Rept. No. SAND-77-1259 CONF-780406-1, 3pp., 1977. Sandia National Laboratories, Albuquerque, NM, USA. Presented at American Welding Society Annual Meeting, 3 April 1978, New Orleans, LA.

The feasibility of welding thin-walled high vacuum devices has been evaluated with the goal of defining a process capable of consistently producing welds which are leak-tight and which possess sufficient fracture toughness so as not to develop leaks later. The principal problem in welding molybdenum is related to a high propensity for intergranular fusion zone cracking due to grain boundary oxide films. The oxygen content of commercially available grades of molybdenum is sufficient to embrittle grain boundaries of typical coarse-grained fusion zones. The overall experimental programme described involves two approaches to improving the ductility of the molybdenum welds: grain refinement, and reduction of the oxygen content of the fusion zone.

4.4.3 (2) Dependence of fracture toughness of molybdenum laser welds on dendritic spacing and in situ titanium additions

Jellison, J. L. and Pope, L. E. (Sandia Laboratories, USA). Rept. No. SAND-79-0193C CONF-790518-6, 12 pp., 1979. Sandia National Laboratories, Albuquerque, NM, USA. Presented at 10th US/UK Symp. on Neutron Generators, 21 May 1979, Albuquerque, NM, USA. Also SAND-80-1905C, 18pp., 1980 and CONF-801132-1, presented at Lasers in Manufacturing Conf., 11 November 1980, Los Angeles, CA.

The fracture toughness of molybdenum welds has been improved by in situ gettering of oxygen by means of physically deposited titanium. The addition of titanium suppressed brittle intergranular fracture. Pulsed laser welds (both Nd:YAG and CO_2) exhibited superior toughness to that of continuous wave CO_2 laser welds.

4.4.3 (3) Mechanical properties, fracture toughnesses and microstructures of laser welds of high strength alloys

Metzbower, E. A. and Moon, D. W. (US Naval Research Laboratory, USA). In, *Proc. Applications of Lasers in Materials Processing Conf.*, 18-20 April 1979, Washington, DC, pp. 83-100. ASM (Materials/Metalwork Technology Series), 1979.

The materials studied were HY-80, HY-130, aluminium and titanium. Hardnesses are related to fast cooling rate, and porosity is a problem which can be alleviated by selection of shielding gases and effective gas coverage. The fine grain size produced in the fusion zone and the narrow heat affected zone contribute to good mechanical properties. Second phase vapourisation lowers inclusions and changes inclusion distribution. (3 refs.)

4.4.3 (4) Monitoring laser welds using stress wave emission (SWE) techniques

Jon, M. C. In, *International Advances in Non-Destructive Testing*, Vol. 6, pp. 351-369. Gordon and Breach Science Publishers Inc, New York, NY, USA, 1979.

Stress wave emission (SWE) techniques are used to monitor the quality of laser welds. From the experiments described, it is shown that the weld mechanism for laser welding of insulated Cu wire to a 0.04in. diameter brass terminal produces a predictable SWE signature when detected by a 1MHz sensor followed by a 0.6-1MHz narrow band filter. Mechanisms for SWE generation during laser welding are also discussed. (10 refs.)

4.4.3 (5) Pulsed YAG laser welding of ODS alloys

Kelly, T. J. (International Nickel Co, USA). In, *Proc. Applications of Lasers in Materials Proces-*

sing Conf., 18-20 April 1979. Washington, DC, pp. 43-50, ASM (Materials/Metalwork Technology Series), 1979.

Describes the welds of Incoloy alloy MA 956 for morphology and examination of distribution of dispersoid by X-ray and transmission electron microscopy. Properly oriented pulsed YAG laser welds have resulted in retention of a large portion of high temperature properties. (3 refs.)

4.4.3 (6) Experimental evaluation and finite element analysis of laser welded copper-copper and copper-aluminium joints

Jones, M. G. and Wang, H. -P. In, *Proc. Advances in Laser Engineering and Applications Conf.*, Vol. 247, 31 July – 1 August 1980, San Diego, CA, pp. 45-54. Society of Photo-Optical Instrumentation Engineers, Bellingham, WA, 1980.

The effect of temperature as a function of time in laser welding of similar and/or dissimilar materials was evaluated experimentally and by finite element analysis. Three cases-Cu-Cu, Cu-Al and Al-Al welded joints were analysed by simulation. Tensile strength, relative impact strength, resistance vs. temperature variation and an electron microprobe analysis were evaluated. Good tensile strength and low electrical resistance results were obtained for welding ETP Cu with a Nd:glass laser. Copper-aluminium conductors responded with good tensile and impact strengths to laser welding. (7 refs.)

4.4.3 (7) Rapid solidification processing of superalloys using high-power lasers

Breinan, E. M., Kear, B. H. and Snow, D. B. In, *Proc. Superalloys 1980 Conf.*, 21-25 September 1980, Champion, PA, pp. 189-203. American Society for Metals, OH, USA, 1980.

Describes the LAYERGLAZE process whereby bulk parts can be built up from rapidly solidified materials. Involves the introduction of wire or powder feed on to a moving substrate at the point of impingement of a continuous CO_2 laser beam. LAYERGLAZE processing is currently being employed to fabricate model gas turbine discs with very fine microstructures and low defect populations. This application has required the development of high-Mo, Ni-base superalloys (8-12-3) which are laser-weldable, phase stable and strong. (7 refs.)

4.4.3 (8) A comparative evaluation of laser and gas tungsten arc weldments in high-temperature titanium alloys

Baeslack, W. A. and Banas, C. M. *Welding J.*, 60(7): 121-130, 1981.

The results of mechanical property tests showed laser beam and gas W arc weldments to exhibit as-welded and postweld heat treated strengths superior to those of the alpha – beta processed base materials. The high-strength, predominantly martensitic microstructures and coarse, prior-beta grain macrostructures which characterised the rapidly-cooled gas W arc weldments combined to promote poor bend and tensile ductilities in the as-welded condition. A considerably smaller prior-beta grain size associated with the laser beam weldment fusion zones contributed to enhanced ductilities at comparable or superior strength levels. Although aging of the martensite during low-temperature post-weld heat treatment further strengthened and embrittled the weldments, intermediate and high-temperature, heat treatments overaged and coarsened the microstructures, resulting in an improvement in fusion zone ductilities at the sacrifice of strengths. As with as-welded properties, the ductilities of heat treated laser beam weldments were found to be superior to those of rapidly-cooled gas W arc weldments, despite exhibiting similar fusion zone strengths and microstructures. (24 refs.)

4.4.3 (9) Fractography of laser welds

Metzbower, E. A. and Moon, D. W. In, *Proc. Fractography and Materials Science Conf.*, 27-28 November 1979, Williamsburg, VA, pp. 131-149. American Society for Testing and Materials, Philadelphia, PA, USA, 1981.

A series of high-strength alloys (HY-80, HY-130, 5456 Al and 6211 Ti) have been laser-beam welded in thicknesses of up to 1.27cm. These welds have been tested for their mechanical properties, fracture toughness and hardness. The microstructures of these weldments have been determined in the base plate, heat-affected zone and fusion area. The fractographic features for each alloy have been determined and correlated with the weldment's properties and structures.

4.4.3 (10) Welding of mechanically alloyed ODS materials

Kelly, T. J. In, *Proc. Conf. on Trends in Welding Research in the United States*, 16-18 November 1981, New Orleans, LA, pp. 471-487. American Society for Metals, OH, USA, 1982.

Pulsed laser welds and electron beam welds of oxide dispersion strengthened Incoloy alloy MA 956 have been examined by transmission electron microscopy to determine the dispersoid morphology and distribution. The results indicate that the dispersoid remains suspended in the alloy without agglomerating or floating out of the fusion zone. Grain boundary formation perpendicular to the directional structure of Incoloy alloy MA 956 can also be altered by use of either the laser or electron beam welding process combined with proper joint design. Due to the strong preferred orientation of the alloy, the fusion zone solidifies epitaxially from the side walls of the weld cavity to within a narrow zone near the weld centreline. This centreline is comprised of equiaxed grains. If the weld cavity is oriented properly, the epitaxial growth will orient the grains parallel to the base metal structure. The only grain boundaries to form perpendicular to the base metal structure will be those equiaxed grains which form at the weld centreline. This condition results in a higher effective weld strength for the lap joint design. (8 refs.)

4.4.3 (11) Microstructure and mechanical properties of laser welded titanium 6Al-4V

Mazumder, J. and Steen, W. M. *Metall. Trans. A*, 13A(5): 865-871, 1982.

Laser butt welds were fabricated using a control 2kW CW CO_2 laser. The relationships between the weld microstructure and mechanical properties are described and compared to the theoretical thermal history of the weld zone as calculated from a three-dimensional heat transfer model of the process. The structure of the weld zone was examined to detect any gross porosity and to identify the microstructure. The oxygen pick-up during laser welding was analysed to correlate further with the observed mechanical properties. It was found that optimally fabricated laser welds have a very good combination of weld microstructure and mechanical properties, ranking this process as one which can produce high-quality welds. (26 refs.)

4.4.3 (12) Microstructures and mechanical properties of laser welded titanium alloys

O'Neal, J. E. et al. In, *Proc. Advanced Processing Methods for Titanium Conf.*, 13-15 October 1981, Louisville, KY, pp. 189-201. Metallurgical Society/AIME, Warrendale, PA, 1982.

Investigates laser weldings of α, $\alpha + \beta$, and β Ti alloys properties. Compares results of laser welding studies with those of electron-beam and gas-tungsten-arc weldings. In addition, the effects of laser power and traverse rate on melt width and melt depth are demonstrated.

4.4.3 (13) Study of the dependence of microsegregation on critical solidification parameters in rapidly-quenched structures

Breinan, E. M. and Snow, D. B. (United Technolo-

gies Research Center, USA). Rept. No. UTRC/ R82-915797-2, 53pp., 1982. United Technologies Research Center, East Hartford, CT, USA, 1982.

Laser welds of 1.27 and 2.54cm thick Ti-6Al-4V plate separated by a 4mm straight sided gap were achieved with the aid of the addition of solution blend Ti-6Al-4V powder to the weld gap. A continuous CO_2 laser was utilised in the unstable resonator mode with a set beam power of 5kW, focused by a copper mirror of 45.7cm focal length. Each successive layer of feedstock filled the gap width, and was overlaid by another until the gap was filled. Ambient temperature tests of cross-weld tensile and impact specimens revealed no fusion zone strength reduction but a somewhat decreased toughness with respect to the annealed base metal.

4.4.3 (14) Comparative evaluation of fatigue endurance of welded titanium alloy joints

Grigoryants, A. G. et al. *Avtom. Svarka*, (4): 16-18, 1983. (In Russian)

Evaluation of welds produced in PT-3V Ti alloy plates by Argon arc, EB and laser techniques. Fatigue endurance limit for laser welds was superior to that of EB and Argon arc weldments. Laser welds also had higher impact toughness and exhibited finer microstructure in the weld metal and HAZ. (7 refs.)

4.4.3 (15) Effect of technical measures on the reheat cracking resistance of welded joints in nickel alloys in heat treatment

Bagdasarov, Yu. S. et al. *Weld. Prod.*, 30(4): 41-44, 1983.

The reheat cracking resistance is determined by the mechanical properties of the heat-affected zone metal. The effect of the residual welding stresses is stronger than that of the stresses generated during the formation of the gamma phase. The high fracture resistance of the welded joints made by laser welding is affected by combining the high mechanical properties of the HAZ with the potential energy of the residual welding stresses. The results show that the resistance of the welded joints to reheat cracking increases in welding with filler materials, resulting in reduced rates of dispersion hardening of the weld metal in relation to the HAZ metal. The fracture resistance of the HAZ metal increases with decreasing time during which the metal is kept in the high-temperature brittleness region. This can be ensured by, for example, welding with forced cooling.

4.4.3 (16) Influence of technological procedures on crack resistance during heat treatment

Baldasarov, Yu. S. et al. *Svar. Proizvod*, (4): 23-26, 1983. (In Russian)

Samples of Ni-based alloy KhN50MVKTYuR were welded by regular Argon arc welding, arc welding with electromagnetic stirring of the metal in weld pool; electron beam welding and laser welding.

All processes except the first significantly improved the properties of the metal in the heat affected zone in comparison with those provided by regular arc welding. The effect of residual welding stresses on crack resistance appeared to be more pronounced than that caused by volume changes at heat treatment due to the formation of gamma phase. It was established that resistance to cracking at heat treatment can be improved by use of additives which retard the precipitation hardening of the welded metal.

4.4.3 (17) The microstructure and mechanical properties of a welded molybdenum alloy

Wadsworth, J. (Lockheed Palo Alto Research Laboratory, USA). *Mater. Sci. & Engng.*, 59(2): 257-273, 1983.

Wrought Ti-Zr-Mo (TZM) alloy has been welded using laser welding techniques and the microstructure, tensile properties and fracture surfaces of these welded samples have been examined. Although the welds have been found to be defect-free, a disparity in grain size leading to large strength differences exists between the weld and parent metal. Tensile tests have revealed that fusion zone strengths are typical of those expected for the grain size in the weld metal. However, brittle behaviour is also always observed, with fracture initiating at grain boundary embrittlement. It is proposed that brittle behaviour is a result of local high strain rates in the weld zone. (36 refs.)

4.4.3 (18) Modification of weld fusion zone grain structure in thorium-doped iridium alloys

David, S. A. and Liu, C. T. In, *Proc. Grain Refinement in Castings and Welds Conf.*, 25-26 October 1982, St. Louis, MO, pp. 249-258. Metallurgical Society/AIME, Warrendale, PA, USA, 1983. Also *Welding J.*, 61(5): 157-163, 1982.

Reviews the arc and laser weldability of two Ir 0.3% W alloys containing 60 and 200wt. ppm Th. These alloys are prone to hot cracking during gas tungsten arc welding, but the alloy can be successfully welded with a continuous wave high-power CO_2 laser system or an electron beam welding system. Successful laser welds without hot cracking have been attributed to the highly concentrated heat source available in these systems and the refinement of the fusion zone grain structure. Microstructure refinement of the fusion zone in these alloys is critically reviewed and discussed in the light of weld puddle geometry, welding speed and the turbulent action within the weld puddle. (13 refs.)

4.4.3 (19) Transverse deformations of AMg6 alloy in laser, electron beam and argon arc welding

Ivanov, V. V. et al. *Avtom. Svarka.*, (11): 21-22, 1983. (In Russian)

It was shown that welding speed strongly affects transverse shrink. In general, higher speed results in lower transverse shrinkage along the seam length. Transverse shrinkage after laser welding is less than that observed after argon arc welding.

4.4.3 (20) Properties of the heat resistant nickel alloy KhN68VMTYuK during beam and argon arc welding

Fedorov, B. M. et al. *Svar. Proizvod.*, (11): 19-21, 1984. (In Russian)

The high temperature, long-term creep strengths and fatigue lives 1.5mm thick joints in KhN68VMTYuK alloy made by laser welding in helium are shown to be higher than those of Ar arc welded joints. Laser welding in air produces joints that are somewhat less ductile and have a lower fatigue limit in short-term tests than welds made in He; the properties of these joints, however, are superior to those of Ar arc welds. It is also shown that the high-temperature short-term mechanical properties of laser welds are improved as the welding speed increases.

4.4.3 (21) Softening of heat-affected zone metal in welding creep-resisting nickel alloys

Yakushin, B. F. and Fedorov, B. M. *Weld. Prod.*, 31(9): 11-13, 1984.

Outlines arguments that the advantage of laser welding dispersion-hardened nickel alloys in comparison with arc welding is the fivefold reduction of the width of the zone of thermal softening and the possibilities of producing the HAZ of welded joints of equal strength to that of the parent metal.

4.4.3 (22) Fracture toughness testing of laser welds in 2mm thick titanium alloy IMI 550

Scott, M. H., Gordon, J. R. and Gittos, M. F. (Welding Institute, UK). Rept. No. REPT-378411185, BR95196, 27pp., March 1985. Welding Institute, Cambridge, UK.

Compact specimens are used to assess toughness in the as-welded state and after a stress-relieving heat treatment. Tensile and hardness tests are also made on the welds. In the as-welded state, the weld metal has similar or greater toughness than the parent metal but heat treatment causes a reduction below that of the as-received parent metal. In all cases, fracture takes place in a brittle manner. The welds are harder than the parent metal with lower proof stresses but higher tensile strengths. Heat treatment raises proof stresses without markedly affecting tensile strength.

4.4.3 (23) Laser cutting of Al 7075 sheets

Dillio, A. (Universita dell'Aquila, Italy). et al. In, *Proc. 2nd Int. Conf. on Lasers in Manufacturing*, 26-28 March 1985, Birmingham, UK, (Ed. M. F. Kimmitt), pp. 57-62. IFS (Publications) Ltd, Bedford, UK, 1985.

This work is concerned with surface morphology and changes in mechanical and structural properties in laser cutting of aluminium alloy sheets. 1.2 and 2mm. thick Al sheets for aircraft construction were cut with a 2.0kW CW CO_2 laser using two different shielding gases: O_2 and He. Surface roughness was measured adopting mechanical holographic techniques on cut surfaces. These tests allow the determination of the most important morphological parameters: roughness index R_a, bearing area Sp, and number N of profile peaks intersected by the mean line. Microhardness tests were conducted in order to evaluate mechanical properties and metallographic tests to identify structural modifications.

4.4.3 (24) Pulsed laser weldability of aluminum alloys

Weeter, L. A. (Monsanto Research Corp, USA). Rept. No. MLM-3320(OP) CONF-8511103-4, 9pp., 1985. Presented at *Laser Welding Machining and Materials Processing – Proc. Int. Conf. on Applications of Lasers and Electro-Optics, ICALEO '85*, 11-14 November 1985, San Francisco, CA, (Ed. C. Albright), pp. 81-88. Laser Institute of America, Toledo, OH, USA/IFS (Publications) Ltd, Bedford, UK, 1986.

This study was undertaken to determine the weldability of six aluminium alloys in similar alloy, dissimilar alloy, and similar alloy with 4047 filler metal addition combinations. The pulsed laser weldability test rated the weldability of the six aluminium alloys on the basis of crack sensitivity. The results of joining 1100, 3003, 5356, or 6061 to either 4043 or 4047 in an approximately 50% mixture revealed that all of these combinantions were very crack sensitive. The addition of smaller amounts of 4047 to either 5356 or 6061 revealed the same phenomenon.

4.4.3 (25) Weldability of copper alloy with YAG laser

Matsunaga, K. *J. Jpn. Soc. Precis. Engrs.*, 51(8): 1563-1568, 1985. (In Japanese)

Welding conditions and their relation to mechanical characteristics and metal structures and examined. The main results are summarised. The welded part of the metal structures is dendrite. This confirms the reliability of the metal structures and hardness characteristics under YAG laser operating conditions.

4.4.3 (26) Laser production of ultra-fine metallic and ceramic particles

Matsunawa, A. and Katayama, S. (Osaka University, Japan). In, *Laser Welding Machining and Materials Processing – Proc. Int. Conf. on Applications of Lasers and Electro-Optics, ICALEO '85*, 11-14 November 1985, San Francisco, CA, (Ed. C. Albright), pp. 205-212. Laser Institute of America, Toledo, OH, USA/IFS (Publications) Ltd, Bedford, UK, 1986.

Since ultra-fine metallic and ceramic particles are expected to be new, promising, high-performance materials, the authors have attempted to produce ultra-fine particles of metals, oxides and nitrides by pulsed Nd:YAG laser irradiation heating-evaporation process of their metallic materials in argon (Ar), helium(He), oxygen(O_2) or nitrogen(N_2) atmosphere, and to reveal the characteristics of their morphology, crystal structure, size distribution, etc. (12 refs.)

4.4.3 (27) Microstructure and wear properties of laser clad Fe-Cr-Mn-C alloys

Singh, J. and Mazumder, J. (University of Illinois, USA). In, *Laser Welding, Machining and Materials Processing – Proc. Int. Conf. on Applications of Lasers and Electro-Optics, ICALEO '85*, 11-14 November 1985, San Francisco, CA, (Ed. C. Albright), pp. 179-186. Laser Institute of America, Toledo, OH, USA/IFS (Publications) Ltd, Bedford, UK, 1986.

In a recent investigation by Eiholzer, et al., it was reported that the wear properties of laser clad Fe-Cr-Mn-C alloy were found to be superior

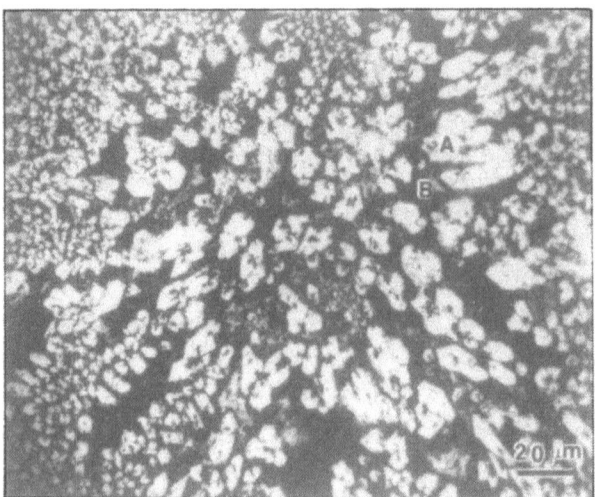

Optical micrograph of laser cladded sample showing
duplex microstructure (3kW, 20in./min)
(See 4.4.3 (27))

to Stellite-6. This provided great incentive for a
detailed study of the transformation product in
the above ternary and quarternary systems by
laser techniques. The present investigation is an
extension of the earlier work and is now focused
on the microstructural changes occurring in
Fe-Cr-Mn-C alloy produced by laser surface

cladding under different processing conditions.
The optimum processing conditions were also
discussed. The microstructure of the samples
were characterised by optical microscopy, SEM,
Auger, and STEM.

4.4.3 (28) A study of the effect of essential variables on high-power laser beam welding of Inconel 600

Bruck, G. J. et al. (Westinghouse Corp, USA). In,
Laser Welding, Machining and Materials Processing – Proc. Int. Conf. on Applications of Lasers
and Electro-Optics, ICALEO '85, 11-14 November
1985, San Francisco, CA, (Ed. C. Albright), pp.
11-18. Laser Institute of America, Toledo, OH,
USA/IFS (Publications) Ltd, Bedford, UK, 1986.

A parametric investigation was conducted to
evaluate the effect of several essential variables
on high-power laser beam welding of Inconel
600. Power level, travel speed and focal point
elevation were systematically varied for flat,
horizontal and vertical position welding. Metallographic examination characterised the effects
in terms of penetration, freedom from defects
and bead geometry.

4.5 LASER PROCESSING SAFETY MEASURES

4.5 (1) Danger of radiation damage to the eyes, especially laser radiation

Fankhauser, F. and Lotmar, W. *Zeitschrift Angewandte Physik.* 6: 521–524, 1966. (In German). (Royal Aircraft Establishment Translation No. 1228).

Examines the effects of laser radiation and recommends filters and goggles designed to protect the eye.

4.5 (2) The biologic effects of ultraviolet radiation (with emphasis on the skin)

Urbach, F. Pergamon Press, UK, *Proc. 1st Int. Conf.*, 1969, 704pp.

Includes some considerations of laser radiation.

4.5 (3) Health hazards from laser material processing

Sliney, D. (US Army Environmental Hygiene Agency, USA). SME Tech. Paper No. MR75-581, 9pp., December 1975. Society of Manufacturing Engineers, Dearborn, MI, USA.

A great deal of experience has evolved in evaluating and controlling the hazards from laser drilling, cutting, scribing, and welding devices. The attendant hazards vary greatly in magnitude from small-scale neodymium IC mask scribers to high-power, CO_2 laser welding and cutting systems. The principal hazards are eye or skin injuries from reflected laser radiation and the potential occupational diseases from contaminants produced in the process. Enclosure is recommended, because it not only contains the radiation but also contains the contaminants. (9 refs.)

4.5 (4) Laboratory safety manual

British Ceramic Research Association, Stoke-on-Trent, UK, 1975, 48pp. (4th Edition)

Includes guidance on the use of lasers.

4.5 (5) Laser hazard classification guide

Sliney, D. H., et al. (National Institute for Occupational Safety, and Health, USA). National Institute for Occupational Safety and Health, Cincinnati, OH, USA, 1976, 206pp.

4.5 (6) Laser safety handbook

Chabot, L. and Mallow, A. Van Nostrand, New York, NY, USA, 1978, 353pp.

A description of the operation and hazards of lasers is followed by chapters on measurement of output and exposure, protective standards, evaluation and classification of hazards, control of hazards, safety legislation, medical surveillance and protective eyewear.

4.5 (7) Requirements concerning the electrical safety of laser-apparatus and installations

European Committee for Electrotechnical, Standardisation (ECES), 1978, 53pp. (CENELEC harmonisation document HD 194)

4.5 (8) Threshold limit values for physical agents

Largent, E. J. (American Conference of Governmental Industrial Hygienists). *National Safety News*, 118(6): 60-68, 1978.

Changes in the 1978 list applying to ionising radiation and microwaves are summarised. Gives TLVs for heat stress and ionising radiation including various kinds of laser. Notice of intent is given to establish TLVs for light and near-infrared radiation.

4.5 (9) How to fit eye protection to the work hazard

National Safety News, 119(4): 46-55, 1979.

Considers selection of eye-wear, safety spectacles and contact lens impact protection by chemical hoods, goggles or face shields and also looks at eye protection for specific hazards such as laser beam welding.

4.5 (10) Radiation control: A decade of progress

Proc. 10th Annual Conf. on Radiation Control: A Decade of Progress, April-May 1978, Harrisburg, PA. Food and Drug Administration HEW Publication 79-8054, 1979, 399pp.

Papers cover non-ionising and ionising radiation, microwaves, lasers, health risks and other issues

4.5 (11) Hazard analysis on Gaussian shaped laser beams

Marshall, W. J. *American Industrial Hygiene Assoc.*, 41(8): 547-551, 1980.

Develops formulae for hazard evaluation based on a Gaussian beam shape. These formulae may be applied to a variety of situations and other formulae are developed for calculating beam diameter, irradiance, nominal ocular hazard distance, atmospheric absorption, and required optical density for laser safety eyewear.

4.5 (12) Health hazards of welding

Doig, A. T. British Safety Council, UK. British Safety Council, UK, 1980, 12pp.

Examines the dangers of lasers to the eyes and skin, the problems of metal fume fever, radiation and poisoning by CO, nitric oxides and mercury.

4.5 (13) Laser safety reference book

Rockwell, R. J. and Sliney, D. H. Laser Institute of America, Cincinnati, OH, USA, 1980.

Reference work which examines US standards and safety legislation, considers the radiation problems from laser beams, and recommends goggles and other forms of eye protection from optical radiation hazards.

4.5 (14) Non-ionising radiation: proceedings of a topical symposium

La Nier, M. E. and Murray, W. E. (Eds.) *American Conference of Governmental Industrial Hygienists,* Cincinnati, OH, USA, 1980, 276pp. Proc. of Symp. 26-28 November 1980, Washington, DC.

Includes consideration of radiation hazards of lasers to the eyes and skin.

4.5 (15) Radiation

Industrial Safety, 26(718): 1-8, 1980. (Industrial Safety Data Sheet No. 20)

Describes the types of radioactive emissions – alpha particles, beta emissions, gamma radiation, Bremsstahlung, neutrons, nuclides, and radio-nuclides. It also lists types of electromagnetic waves including lasers, describing their uses and potential adverse effects. New and old radiation units are explained, as is radioactive decay. The legislation is outlined as related to the relevant acts and regulations, on use in the workplace.

4.5 (16) Safety with lasers and other optical sources: A comprehensive handbook

Sliney, D. and Wolbarsht, M. Plenum Press, New York, NY, USA, 1980, 1035pp.

Comprehensive text book which covers safety practices in the use of lasers in a wide range of applications including welding. Radiation hazards, skin and eye protection, medical examination, etc. are all included.

4.5 (17) Trends in welding safety

Smart, R. British Federal Welder Machine Co, Ltd, UK, 1980, 102pp.

Includes advice on safety in laser welding operations.

4.5 (18) Evolution of laser machining and welding, with safety

Charschan, S. S. (West Electric Co, USA). In, *Proc. SPIE, Ocular Effects of Non-Ionising Radiation Conf.,* Vol. 229, 7 April 1981, Washington, DC, pp. 144-153. Society of Photo Optical Instrumentation Engineers, Bellingham, WA, USA, 1980.

Current implementation of the laser provides economical technology. In addition, components are now being designed so that they can only be processed by a laser. Describes the development of laser material processing applications with emphasis on machining and welding and the associated safety measures. (16 refs.)

4.5 (19) Lasers

Danielson, G. (Arbetarskyddsstyrelsen, Sweden). Liber Tryck, Stockholm, Sweden, 1981, 28pp. (AFS 1981:9) (HSE Translation No. 9799)

Examines Swedish safety regulations and protection from laser radiation. Assesses various permissible exposures and biological effects of lasers.

4.5 (20) Eye protection brought into perspective

Farmer, D. *Health and Safety at Work,* 3(8): 20-24, 1982.

Includes brief comment on lasers and welding as processes from which eyes are at risk. Looks at accident statistics in the field plus current legislation and regulations. Comments on the fact that in the majority of cases, eye protection had been provided but not used. Also covers problems which arise concerning contact lenses and the doubts about providing prescription safety glasses.

4.5 (21) Laser in working life: safety aspects

Stenow, A. *Safety Management*, 8(6): 34-36, 1982.

Considers dangers of laser radiation and discusses laser protection problems. Covers characteristics of laser emission and damage and the injury it can cause. Lists a variety of laser applications in working life and discusses their safety aspects.

4.5 (22) Lasers and optical radiation

United Nations Environment Programme, World Health Organisation Geneva, Switzerland, 1982, 154pp.

Makes recommendations on eye and skin protection, exposure limits, protection from welding arcs and accident procedures. Also assesses medical surveillance and training in safety procedures.

4.5 (23) Lasers in construction

Construction Safety Association of Ontario, Canada, 1982, 1p. (Information bulletin IBI).

Looks briefly at protection of eyes for laser beams and radiation.

4.5 (24) Non-ionising radiation protection

Suess, M. J. World Health Organisation, Copenhagen, Netherlands, 1982, 267pp.

Includes consideration of radiation protection when using lasers.

4.5 (25) Occupational eye injuries through lasers

Von Felixberger, F. and Szadkowski, D. *Arbeitsmedizin, Sozialmedizin, Praeventivmedizin*, 17(1): 14-17, 1982. (In German).

Reviews the incidence of laser eye injuries and gives a summary of the clinical picture of them. Emphasises the need for effective safety procedures.

4.5 (26) Site safety

Laney, J. C. Construction Press, 1982, 218pp. (Site practice series).

Book covers a wide range of safety practices on construction sites including safety in the use of laser welding equipment.

4.5 (27) Biological bases of the maximum permissible exposure levels of the UK laser Standard BS 4803: 1983

McKinlay, A. F. and Harlen, F. National Radiological Protection Board, Chilton, UK, 1983, 49pp. (NRPB R153)

4.5 (28) Guidelines on health hazards of electromagnetic radiation

Health and Safety Bulletin, (31): 1-88, 1983.

Lists sources of exposure, occupations at risk, health hazards of microwave and radiofrequency radiation, visible light, including lasers, ultraviolet and ionising radiation. Describes the standards and regulations applying to electromagnetic radiation in Australia and lists the major surveys of EMR levels which have been completed in Australia. Outlines the proposed Australian Council of Trade Unions' (ACTU) health and safety policy for electromagnetic radiation.

4.5 (29) Health and safety in welding and allied processes

Balchin, N. C. Welding Institute, Cambridge, UK, 1983, 198pp. (3rd edition)

4.5 (30) High-tech safety at Westinghouse

Occupational Hazards, 45(10): 76-77, 1983.

Describes how Westinghouse has developed safety practices for robots, lasers and advanced electronic equipment.

4.5 (31) Laser injuries

Occupational Health and Safety, p. 7, January 1983.

Brief comment on the accident where a research chemist lost much of the visual acuity in one eye, after being hit by a reflected beam while adjusting the frequency doubler on his dye laser.

4.5 (32) Laser safety: The hazards in perspective

Henderson, A. R. (Cambridge Consultants Ltd, UK). In, *Proc. 1st Int. Conf. on Lasers in Manufacturing*, 1-3 November 1983, Brighton, UK, (Ed. M. F. Kimmitt), pp. 193-204. IFS (Publications) Ltd, Bedford, UK, 1983.

The safe use of lasers requires an understanding and appreciation of the hazards associated with their use, and a knowledge of the means by which these hazards can be controlled. Unfortu-

nately, laser safety codes are of necessity rather complex, since lasers cannot be regarded as a single group to which a common exposure limit can be applied. It is the intention of this paper to explain the reasons for this to describe the hazard classification scheme for lasers, and to discuss current thinking on management and engineering controls, and on the personnel protection procedures necessary to ensure that lasers can be safely used in an industrial environment.

4.5 (33) New radiation safety controls in universities

Health and Safety Information Bulletin, (90): 2-5, 1983.

Discusses the implications for proposed ionising radiation controls for universities and possible problems over using existing radiation protection officers as 'Radiation Protection Advisors'. Discusses also the British Standard on laser safety, BS 4803: 1983.

4.5 (34) Optical radiation hazards of laser welding processes. Part 1 Neodymium – YAG laser

Rockwell, R. J. and Moss, C. E. *Am. Ind. Hygiene Assoc. J.*, 44(8): 572-579, 1983.

There are numerous industrial laser applications where systems are required to be used in an unenclosed manner. In such applications, there is concern for both eyes and skin hazards caused by direct and scattered laser radiation, as well as potential hazards caused by the optical radiation created by the laser beam's plume radiation. Radiant energy measurements are reported for both the scattered laser radiation and the resultant plume radiations which were produced during typical unenclosed Nd:YAG laser welding processes. The data obtained were compared to applicable safety standards.

4.5 (35) Personal eye protection: eye protectors for adjustment work on lasers and laser systems. Laser adjustment eye-protectors

Institut Belge de Normalisation, Brussels, Belgium, 1983, 9pp. (HSE Trans. No. 10662).

4.5 (36) Personal eye protection. Filters and eye protectors against laser radiation

(Institut Belge de Normalisation, Brussels, Belgium, 1983, 15pp. (HSE Trans. No. 10661)

4.5 (37) Safety in the use of lasers on site

Cox, E. A. Chartered Institute of Building, Ascot, UK, 1983, 6pp. (Technical information service no. 22).

Covers Helium-neon, Gallium-arsenide, and Nd:glass lasers legislation, hazards, accessible emission limits, maximum permissible exposure and training.

4.5 (38) Scientific and safety aspects of non-ionising radiation. Part 2 lasers

Safety Practitioner, 1(11): 25-27, 1983.

Both direct laser beams and reflections can harm the eye. Discusses the provisions of BS 4803 regarding laser safety and also the types of eye protectors which can be used when working with lasers.

4.5 (39) Some laser safety guidance: but what about by-products

Fletcher, M. J. *Welding and Metal Fabrication*, 51(9): 467-468, 1983.

Mentions that the only legal requirements in the UK and incorporated in the Protection of Eyes Regulations 1974 and the most important safety aspect is that associated with ocular injuries. Screening and the use of goggles are described as well as the alleged hazard from the wearing of contact lenses.

4.5 (40) Adverse health effects of occupational exposure to non-ionising radiation: recent published research

Price, C. (Department of Health and Social Security, London, UK, 1984.

4.5 (41) A functional approach to the assessment of ocular hazards of low-power lasers

Raybourn, M. S. and Kong, R. L. *Health Physics*, 46(1): 107-114, 1984.

An electrophysiological bioassay system has been developed to provide safety-related information about the functional consequences of non-thermal laser irradiation of the vertebrate retina. This preparation allows precise dosimetry at the retinal surface, provides for an intra-retinal control, comparing laser vs non-laser affected tissue, and permits quantitative assessment of the biophysical mechanisms of laser-induced retinal dysfunction.

4.5 (42) Guideline protects laser users

Winburn, D. C. *National Safety News*, 129(5): 51-54, 1984.

Discusses why eye injuries due to industrial lasers are not common in the USA and argues that it is due to safety education, engineering and enforcement of a laser safety policy.

4.5 (43) High technology safety: meeting tomorrow's challenge today

Wilcher, F. E. *Occupational Hazards*, 46(10): 21-22, 25, 1984.

Argues that basic safety principles can be applied to lasers, robot safety and safety override equipment. Also considers safety in terms of motivation of employees.

4.5 (44) Introduction to the British Standard for laser safety and the related biological hazards

Cox, E. A. (Sira Institute, UK). Summary of paper presented at Laser Safety in Industry and the Laboratory Seminar, 20 October 1984, London, UK, pp. 2/1-2/11.

Outlines the potential hazards of lasers to human eyes and skin. Points out that there is no UK legislation setting out specific requirements for the safe use of lasers but British Standard BS 4803: 83(1) provides information and recommendations to assist manufacturers, suppliers and users of laser products to comply with their legal responsibilities imposed by the Health and Safety at Work etc. Act 1974.

4.5 (45) Laser protective eyewear: determining the correct lens

Edmunds, D. *Occupational Health and Safety*, pp. 31-34, July/August 1984.

Discusses the criteria which should be used to assess the type of eye protection required for the type of laser being used.

4.5 (46) Laser safety in industry and the laboratory

Sira Institute, London, UK, 1984. Abstracts of papers presented at Laser Safety in Industry and the Laboratory Seminar, 2 October 1984, London, UK.

Papers cover safety in uses of laser technology in construction, surveying, printing, entertainment and engineering.

4.5 (47) Lasers and occupational health and safety

Cox, E. A. (Health and Safety Executive, UK). *Lasers and Optics International*, 1(1):9, 1984.

Discusses British Standard BS 4803:1983 'Radiation safety of laser products and systems'. The standard gives definitions and information on biological effects of laser radiation; describes the classification of levels of laser radiation, examines hazards of laser products and describes the engineering controls involved in their production, and finally specifies maximum permissible exposure levels.

4.5 (48) Protection against ultra-violet light radiation

Schreiber, P. and Ott, G. (Bundesanstalt fur Arbeitsschutz Dortmund, W. Germany, 1984, 155pp. (Sonderschrift no. 14).

Examines methods of protection of eyes and skin from uv laser radiation.

4.5 (49) Radiation safety of laser products, equipment classification, requirements and user's guide

International Electrotechnical Commission, Geneva, Switzerland, 1984, 137pp. (IEC Publication 825).

4.5 (50) The role of the health physicist in laser safety

Vulpetti, A. T. *Health Physics*, 47(1): 193, 1984. Abstract from papers presented at 29th Annual Meeting of the Health Physics Society, 3-8 June 1984, New Orleans, LA.

An overview is presented of the minimum guidelines for the control of laser hazards including laser and laser system classification, biological effects, control measures, eye protection, laser calculations and measurements and non-radiation hazards of lasers.

4.5 (51) Using lasers – why the macho image worries scientists

Biasio, S. D. *Safety and Fire News*, p. 7, December 1984.

Reviews a conference at which Dr. Gorham is concerned that the public has a distorted image of lasers, and at which the new British Standard on lasers, BS 4803, is summarised.

4.5 (52) Guidelines on limits of exposure to laser radiation of wavelengths between 180nm and 1mm

(International Radiation Protection Association) *Health Physics*, 49(2): 341-359; 1985.

Presents guidelines for protection against

optical radiation emitted by lasers, and argues that with the increase in applications and use of lasers in industry the importance of laser exposure limits will increase. Emphasises the growing need for codes of practice for all potentially hazardous laser operations including materials processing.

4.5 (53) Hazard analysis technique for multiple wavelength lasers

Lyon, T. L. *Health Physics*, 49(2): 221-226, 1985.

Describes a technique for evaluating the hazards from multiple wavelength lasers.

4.5 (54) International symposium on occupational exposure limits

Proc. Int. Symp. on Occupational Exposure Limits, 16-19 April 1985, Copenhagen, NL. American Conference of Governmental Industrial Hygienists, World Health Organisation, Cincinnati, OH, USA, 1985.

Includes papers on laser radiation effects and limits with regard to the human body.

4.5 (55) Laser safety in perspective

Cox, E. A. In, *Proc. 2nd Int. Conf. on Lasers in Manufacturing,* 26-28 March 1985, Birmingham, UK, (Ed. M. F. Kimmitt), pp. 225-236. IFS (Publications) Ltd, Bedford, UK, 1985.

Reviews briefly the range of hazards, the standards and the safety legislation which together establish both the appropriate engineering controls which should be built into the laser product by the manufacturer and the control procedures to be instituted by the laser user to secure its safe use.

4.5 (56) Lasers

Everley, M. *Safety Representative*, p. 5, February 1985.

A summary of safety precautions to be adopted when using lasers particularly in relation to eye protection from radiation.

4.5 (57) Living with new technology: basic laser safety for the safety practitioner

Miles, M. In, *Proc. 3rd ROSPA Int. Safety Exhibition,* 21-23 May 1985. Royal Society for the Prevention of Accidents, 1985.

Reviews the law and codes of practice applicable to lasers. Describes tthe hazardous effects of laser radiation on the human body and the

ancilliary risks from the use of laser apparatus. Discusses practical ways in which risks can be overcome or reduced.

4.5 (58) Occupational hazards from non-ionising electromagnetic radiation

International Radiation Protection Association and International Labour Organisation, Geneva, Switzerland, 1985, 133pp. (Occupational Safety and Health Series no. 53).

4.5 (59) You can live with lasers

Elza, D. *National Safety News*, 131(5): 49-51, 1985.

Discusses the use of lasers in production engineering and steps to take to ensure that employees are protected and that the equipment is correctly maintained.

4.5 (60) Hazardous by-products of plastics processing with CO_2 lasers

Doyle, D. J. and Kokosa, J. M. (GMI Engineering & Management Institute, USA). In, *Laser Welding, Machining and Materials Processing – Proc. Int. Conf. on Applications of Lasers and Electro-Optics, ICALEO '85,* 11-14 November 1985, San Francisco, CA, (Ed. C. Albright), pp. 201-204. Laser Institute of America, Toledo, OH, USA/IFS (Publications) Ltd, Bedford, UK, 1986.

Analysis of the by-products produced by laser cutting of polymethyl methacrylate and polyvinyl chloride indicates the presence of polycyclic aromatic hydrocarbons and other chemicals which are potentially hazardous. Adequate ventilation and monitoring measures should be taken to avoid contact with these materials in the work environment.

4.5 (61) Robot mounted lasers – the safety options

Ward, G. R. and Bandle, A. M. (Health and Safety Executive, UK). In, *Proc. 3rd Int. Conf. on Lasers in Manufacturing,* 3-5 June 1986, Paris, France, (Ed. A. Quenzer), pp. 329-336. IFS (Publications) Ltd, Bedford, UK, 1986.

Robot mounted lasers present a potentially hazardous combination with the possibility of aberrant robot motion coupled with the destructive power of an incorrectly applied laser beam. This paper discusses the safety problems arising from the multi-axis application of lasers afforded by robot mounting and of operating at increasingly higher power levels. The need for extended beam transport systems, rapid switch-

ing from workstation to workstation, continuous wave to pulsed mode operation of the laser and possible developments in automated output focusing, pose a serious challenge to high precision engineering and rapid response control system design. The advantages and disadvantages of various safety strategies such as enclosure with interlocking, active detection scanning systems, high integrity control systems and hybrid systems are considered.

5

INDUSTRIAL APPLICATIONS OF LASERS

5.1 INDUSTRIAL AND AEROSPACE APPLICATIONS

5.1 (1) CNC laser welding pays for itself quickly

Assem. Engng., 19(10): 26-27, 1976.

The unique non-contact nature of laser welding and the ability to join both similar and dissimilar metals with a minimal heat affected zone is proving very popular in many assembly operations. Laser welding has enabled one company, Fischer Porter Co, of Warminster, PA, to simplify product designs and to salvage previously unrepairable expensive subassemblies for their line of process control systems. The cost savings are such that this numerically controlled laser welding system will pay for itself in 15 months.

5.1 (2) Laser cutting for aircraft manufacturing

Huber, J. (Grumman Aerospace Corp, USA). SME Tech. Paper MR76-192, 28pp., Society of Manufacturing Engineers, Dearborn, MI, USA. Presented at Westec Conf., 8-11 March 1976, Los Angeles, CA.

The use of a 250W carbon dioxide laser to rough-cut aircraft structural parts has been in production at Grumman Aerospace Corporation since 1971. Discusses the equipment modifications and tooling requirements to adapt this process to production for both flat and formed parts. The impact of higher laser power levels on cost and performance is evaluated. Typical applications to aircraft parts are shown to illustrate the range of applications.

5.1 (3) Laser cutting of helicopter components

Weld. Met. Fabr., 44(5): 371-372, 1976.

5.1 (4) Laser welding in mass production at Moulinex

Newman, R. A. A. *Weld. & Met. Fabr., 44(8): 563-569, 1976.*

Advantages of using laser beam butt welding for mass production of tins in the food industry are shown. Gives brief details of techniques and metallurgical effects on the metals.

5.1 (5) Production engineering effort on M42/M46 grenade bodies

Tereshkow, H. (AVCO Systems Division, USA). Final Rept. No. AVSD-0318-77-RR, 93pp., November 1977. AVCO Systems Division, Wilmington, MA, USA.

Contains the findings and conclusions resulting from an effort to determine alternative methods for producing the M42/M46 grenade body, using laser techniques to produce lower cost grenade bodies.

5.1 (6) Welding in the aerospace industry – design, materials, welding methods, maintenance

Proc. of Conf., 7-8 December 1978, Berlin, Germany. Deutscher Verlag fur Schweisstechnik (DVS) GmbH, Dusseldorf, Germany, 1978, 134pp.

5.1 (7) Applications of lasers in dental technology

van Benthem, H. and Vahl, J. (University of Muenster, Germany). In, *Conf. Proc. – Laser '79, Opto. Electronics*, 2-6 July 1979, Munich, Germany, pp. 368-373. IPC Science and Technology Press Ltd, Guildford, UK, 1979. (In German)

Discusses laser welding of metallic dental materials. The technique offers the possibility of supplementing and improving the conventional methods of joining used in dental technology. Considers various data on laser parameters that must be considered. (5 refs.)

5.1 (8) Development of laser welding techniques for joining armor

Seaman, F. D. (ITT Research Institute, USA). Quarterly Rept. No. IITRI-B-6157-Q2, 24pp., March 1979. ITT Research Institute, Chicago, IL, USA. See also previous Rept. No. IITRI-B6157-Q1, 26pp., December 1978.

Recently, the laboratory feasibility of heavy-penetration laser welding was demonstrated in low strength steel and the programme describes the establishment of laser welding process parameters that will produce a sound, cost-effective joint with ballistic potential of 1½in. armour plate.

5.1 (9) An economic and technical study on the feasibility of using advanced joining techniques in constructing critical naval marine structures

Rogalsk, W. J. (US Naval Postgraduate School, USA). Rept. No. AD-A092299/7, 184pp., June 1979, US Naval Postgraduate School, Monterey, CA, USA.

A design analysis of a submarine's pressure hull components, fabricated from HY-130 steel, was performed to determine the welding requirements for fabrication. Presented experiments consisting of single- and double-pass laser welding of restrained butt welds in 1in. thick plates. Penetration capabilities obtainable with 12kW of laser beam power were determined. Temperature distributions, longitudinal strains and transverse strains experienced during laser welding of the HY-130 plates were presented in graphical and tabular form. An economic analysis contrasting labour and overheads, filler metal costs and power consumption – comparing shielded metal arc, gas metal arc, laser and electron beam welding processes for fabricating the HY-130 pressure hull was performed.

5.1 (10) Laser drilling for materials fabrication

Heglin, L. M. In, *Proc. Conf. on Applications of Lasers in Materials Processing*, 18-20 April 1979, Washington, DC. pp. 101-119. American Society for Metals, OH, USA, 1979.

The application is a small hole drilling process which is used extensively in the manufacture of aircraft engine components. The principal advantage of lasers is the positioning and automatic cycling capability of the machines which makes it possible to drill diversified hole patterns, with a multiplicity of entrance angles, in one unit. The automated process controls also enhance pattern reproducibility and productivity.

5.1 (11) Laser drilling wins out

Schaffer, G. *Am. Mach.*, 124(8): 112-114, 1980.

Laser drilling was chosen to produce small (0.04-0.50in diameter) cooling holes in the combustor lining in an industrial turbine. Approximately 9000 holes were drilled in each combustor lining. This was complicated by the geometry of the lining and the use of Hastelloy X as the lining material. Because of laser processing's economics and the versatility in using the laser for performing other operations it was s successful drilling method.

5.1 (12) Production laser cutting of gas turbine components

Benedict, G. F. (AiRes Manufacturing Co. of Arizona, USA). SME Tech. Paper No. MR80-851, 13pp., 1980. Society of Manufacturing Engineers, Dearborn, MI, USA.

A large percentage of the components in gas turbine engines are made of sheet metal and require intricate cutting or drilling operations. A 5-axis Nd:YAG laser machining centre was used to provide a more cost effective way of producing these parts.

5.1 (13) Circuit surgery using xenon and YAG lasers

Ravich, G. N. and Waters, R. L. In, *ISTFA 1982 – Proc. Testing and Failure Analysis Conf.*, 25-27 October 1982, San Jose, CA, pp. 86-91. International Society for Testing and Failure Analysis (ISTFA), Torrance, CA, USA, 1982.

Discusses several related applications of lasers manufactured by Florod Corporation. These applications include circuit isolation for internal combustion failure analysis, circuit repair by opening shorts, resistor trimming of functional hybrid circuits and fine Al wire welding.

5.1 (14) CO₂ laser welding of aluminium air spacers for insulated windows

Eckersley, J. S. In, *ICALEO '82, Proc. Materials Processing Symp.*, Vol. 31, 20-23 September 1982, Boston, MA, pp. 61-64. Laser Institute of America, Toledo, OH, USA, 1982.

Industry required better products to provide the possibility of automated handling of air spacer frames. An industrial CO_2 laser was developed to weld the butted edges of the roll-formed shapes at high speeds.

5.1 (15) The high-powered laser welds carbon steel plate

Irving, R. R. *Iron Age*, 225(23): 48-49, 1982.

Describes the experience with laser welding at the Northern Ordnance Division of FMC Corporation a Navy-owned plant fabricating guided-missile launch systems and gun mounts for the US Navy. An HPL series 200 continuous wave CO_2 laser capable of delivering 15kW to the workpiece is being used. A 300% productivity increase, huge cost reduction, lower tooling costs and tougher welds in A36 plates are attained.

5.1 (16) Laser and punch press make a big hit

Albert, M. *Mod. Mach. Shop.* Jan. 1982, 54 (8): pp. 68-74

A sheet metal shop fabricating truck cab components from 14 gauge cold rolled stainless steel uses a laser cutting tool in conjunction with an NC turret punch press.

5.1 (17) Laser precision small hole drilling

Terrell, N. E. *Manuf. Engng.*, 88(5): 76-77, 1982.

Describes the drilling of small holes in a Hastelloy-X liner for gas combustion turbines. Over 9000 holes are generated, ranging in size from 0.046 to 0.049in diameter most of which enter the workpiece at a 25° angle.

It was possible to maintain parallelism of hole sides, eliminate slag at entrance and exit points, reduce the remolten layer and cut drilling time from 14 to 4s.

5.1 (18) Machining aerospace alloys with the aid of a 15kW laser

Hill, V. L., Plankenhorn, D. J. and Rajagopal, S. *J. Appl. Metalwork*, 2(3): 170-184, 1982.

Argues that a laser beam will selectively heat the workpiece in a machining operation and will thereby enable higher rates of metal removal to be achieved without a corresponding loss in tool life. The concept feasibility study involved two aerospace materials: Inconel 718 and Ti-6Al-4V. Ceramic tools were used for Inconel 718 and carbides for the Ti alloy. For both materials, the metal removal rate was significantly increased without aggravating either cutting force or tool wear. The commercial feasibility of laser-assisted machining (LAM) requires incorporation of the process into a laser integrated manufacturing system or dropping the power requirement to approx 5kW through improved beam coupling techniques, both of which serve to reduce the effective capital investment and operating costs for LAM.

Nd: YAG laser drilling machine in operation at Rolls-Royce on production components (See 5.1 (20))

5.1 (19) Laser application aids choice of product design

Anthony, P. (Rofin-Sinar Laser UK Ltd, UK). In, *Proc. 1st Int. Conf. on Lasers in Manufacturing*, 1-3 November 1983, Brighton, UK, (Ed. M. F. Kimmitt), pp. 243-252. IFS (Publications) Ltd, Bedford, UK, 1983.

Holloid Plastics Ltd's decision to produce a high quality vehicle registration plate in acrylic as an injection-moulded component was taken as a result of studies into the effective cost saving and benefits of laser cutting in a continuous production system. The resultant facility breaks away from traditional methods of producing from sheet material. Economic justification of the automatic process is a direct result of incorporation of laser cutting to degate moulded components without any further finishing operation being required. The inherent characteristics of laser cutting, being non-contact and swarf-free, allowed for the viable use of 'pick and place' robotics for automatic handling between the mould and degating station, giving a scratch-free, high quality product.

5.1 (20) Laser drilling of aero engine components

Corfe, A. G. (Rolls-Royce Ltd, UK). In, *Proc. 1st Int. Conf. on Lasers in Manufacturing*, 1-3 November 1983, Brighton, UK, (Ed. M. F. Kimmitt), pp. 31-40. IFS (Publications) Ltd, Bedford, UK, 1983.

The pulsed Nd:YAG laser offers significant cost advantages over other machining techniques for drilling small cooling holes in aero engine components such as combustion chambers and nozzle guide vanes. Recent improvements in laser over E.D.M. have led to the installation into Rolls-Royce at Bristol of a British-built Nd:YAG laser drilling machine totally integrated into a multi-axis CNC system. Rapid drilling rates with good metallurgical and geometric quality have demonstrated the flexibility and potential of the technique and have led to proposals for other machines, either as stand-alone units or as integrated into flexible manufacturing systems.

5.1 (21) Laser-welded blade is key to trenching update

Herbert, S. *Ind. Diamond Rev.*, 43(499): 293-296, 1983. Also *Highways & Public Works*, 51(1884): 18-20, 1983.

High-intensity laser welding has produced a diamond blade which could revolutionise many aspects of highway and construction site cutting, including asphalt, concrete, reinforced concrete,

masonry and roof tiling, as well as stone and refractories. The new laser welded diamond blade can be used with or without coolant, thus giving diamond blade cutting efficiency on a portable basis. Some 16km of asphalt sawing was performed in a recent gas pipelaying project in North-East England.

5.1 (22) Lightweight solar array blanket tooling, laser welding and cover process technology

Dillard, P. A. (Lockheed Missiles and Space Co., Inc, USA). Final Rept. No. NAS 1.26: 170209, LMSC-D843530, 51pp., January 1983. Lockheed Missiles and Space Co, Inc, Sunnyvale, CA, USA.

A two-phase technology investigation was performed to demonstrate effective methods for integrating 50µm thin solar cells into ultra-lightweight module designs. During the first phase, innovative tooling was developed and during the second phase, the tooling was improved and the feasibility of laser processing of lightweight arrays was confirmed. The development of the cell/interconnect registration tool and interconnect bonding by laser welding is described.

5.1 (23) A review of techniques available for girth welding of HSLA pipelines

Hart, P. H. M. et al. In, *Proc. Steels for Line Pipe and Pipeline Fittings Conf.*, 21-23 October 1981, London, UK, pp. 291–301. Metals Society, London, UK, 1983.

The systems available are reviewed and the potential for their application in the next decade is discussed. These include manual metal arc welding with basic coated electrodes, MIG welding and single shot processes such as flash, MIAB, friction, electron beam and laser welding. (26 refs.)

5.1 (24) YAG laser machining center

Schwob, H. P. (Lasag AG, Switzerland). In, *Proc. SPIE Industrial Applications of Lasers Technology Conf.*, Vol. 398, Geneva, Switzerland, pp. 366-373. International Society of Optical Engineers, USA, 1983.

The different methods of laser drilling of cooling holes in jet engine parts are presented.

5.1 (25) CNC laser drilling machine for the production of high quality holes in aerospace components

Arie, G. (J. K. Lasers, France). In, *Proc. Opto 84, 4th Optoelectronic Meeting*, 15-17 May 1984,

Paris, France, pp. 60-62. ESI Publications, Paris, France, 1984. (In French).

In order to meet the needs of high precision hole drilling in gas turbine aircraft engines J. K. Lasers has developed a seven axes CNC automatic drilling machine equipped with a 300W average power Nd:YAG pulsed laser. In addition, the firm has developed a trepanning technique which offers high flexibility.

5.1 (26) Cutting stamping dies with a laser

Gettelman, K. *Mod. Mach. Shop*, 56(8): 50-58, 1984.

By combining a CO_2 laser with a Bridgeport computer numerical control (CNC) milling machine, Photon Sources, Inc, provides the ability to cut punch, die, stripper and guide components economically and rapidly. The laser die approach can be suitable for 80% of the stamping work.

5.1 (27) The design of a CNC laser drilling machine for the production of high-quality holes in aerospace components

Weedon, T. M. (J. K. Lasers Ltd, UK). In, *ICALEO '83, Proc. Materials Processing Symp.*, Vol. 38, 14-17 November 1983, Los Angeles, CA, (Ed. E. A. Metzbower), pp. 192-198. Laser Institute of America, Toledo, OH, USA, 1984.

In the manufacture of gas turbine engines large numbers of holes have to be drilled to allow the flow of cooling air over the most severely thermally loaded components. Such parts vary from small blades to complete combustion assemblies 40in. in diameter and 36in. long. Other aerospace components contain similar holes and may be even larger but they are usually less complex. The holes to be drilled in production components are currently all circular, in the diameter range 0.010 to 1.0in., with the vast majority <0.100in. The adoption of improved equipment means that hole quality and throughput are improved dramatically. The more critical aspects of the design of a successful machine for the production of high-quality holes in gas turbine parts of Hastelloy X Ni-base superalloy are highlighted.

5.1 (28) Laser cutting — a new tool for shipbuilding

Martyr, D. R. (British Shipbuilder, UK). In, *Laser Welding, Cutting and Surface Treatment*, pp. 28-32. Welding Institute, Cambridge, UK, 1984. Also in, *Proc. Developments and Innovations for Improved Welding Production, 1st Int. Conf.*,

13-15 September 1983, Birmingham, UK, pp. 3.1-3.7. Welding Institute, Cambridge, UK, 1983.

Laser cutting is characterised by its ability to produce a clean edge with very low heat input, at high speeds. Describes the way in which a laser cutting system was designed, evaluated, and subsequently built for large steel plate profile cutting in a small warship-building facility, to improve quality and to reduce production costs.

5.1 (29) Laser manufacturing of medical devices

Feeley, J. T. and Hartung, P. *Laser Appl.*, 3(6): 105-108, 1984.

Processes incorporating laser devices include heat treatment, cutting, welding and drilling. A laser-cut Ti wire mesh pad for knee prosthesis is replacing bone cement. Microsurgical instruments fabricated from stainless steel have used laser welding. Nozzles, with openings <0.030in. diameter, can be fabricated from difficult-to-machine materials by laser drilling. (8 refs.)

5.1 (30) Laser welding of drive shafts

Kozlov, A. E. et al. *Avtom. Prom-St.*, (7): 26-27, 1984. (In Russian)

Describes the laser welding of Cardan drive shafts, and the properties of the welded joints of low-carbon steel are examined. Torsion and fatigue test results for laser and arc welded shafts indicate that the mechanical properties of the former are equivalent to or better than those of the arc welded ones.

5.1 (31) CAD-CAM laser cutting of saw blades

Billhardt, C. F. and Wittkopp, C. *Werkstatt und Betr.*, 118(5): 264-266, 1985. (In German)

A previous article by the authors in 1985 dealt with the development work for the programming system, as well as for linking computer aided design with the programming and CNC manufacture. Describes the machine, together with the organisational steps that have been taken with respect to programming, determination of parameters and economic viability.

5.1 (32) CNC and laser open new opportunities

West, B. *Mod. Mach. Shop*, 57(11): 92-96, 1985.

Reports on the installation and implementation of CNC laser, at the Peerless Saw Co. of Groveport, Ohio, for cutting band and circular saw profile from suitable flat stock.

5.1 (33) Improved turbopump dynamics

Kiefling, L. (National Aeronautics and Space Administration, USA). Final Rept. No. NAS 1.15: 86514, 6pp., July 1985. National Aeronautics and Space Administration, Huntsville, AL, USA.

A study was initiated to investigate the practicality of increasing rotor critical speeds by changes in manufacturing method. The aim was to build a pump with an all laser welded shaft and case; the unit to be opened by laser cutting and rebuilt by rewelding the same surface. Use of a split casing, common in industry, would permit assembly of the rotor outside the case.

5.1. (34) Laser beam replaces plasma arc welding

Weld. Des. & Fabr., 58(8): 42-45, 1985.

By using laser beam welding, author claims a 90% reduction in production cost of jet-engine air impingement tubes.

5.1 (35) Laser machining of aero engine components

Baker, G. (Amchem Co, UK). *Aircr Engng.*, 57(2): 15-16, 1985.

Based on substantial in-house experience with the development and use of laser drilling systems for sub-contract work, Amchem has recently produced a laser machining centre designed primarily for use on combustion chambers. Using a YAG laser, it avoids many of the problems of conventional, violent laser drilling processes, giving smoother surface finishes, and better control.

5.1. (36) Welding with high power CO_2 lasers

Bakowsky, L. *Elektro – Anz*, 38(11): 43-44, 1985. (In German)

Since the establishment of laser cutting as a flexible manufacturing method, laser welding is being introduced into mass production, and is particuarly suitable for engine and gear manufacture.

5.1 (37) Systems considerations for multiaxis CO_2 laser material processing

VanderWert, T. L. (Data Card Corp, USA). In, *Laser Welding, Machining and Materials Processing – Proc. Int. Conf. on Applications of Lasers and Electro-Optics, ICALEO '85*, 11-14 November 1985, San Francisco, CA, (Ed. C. Albright), pp. 101-106. Laser Institute of America, Toledo, OH, USA/IFS (Publications) Ltd, Bedford, UK, 1986.

Major productivity increases have been reported by the growing number of users of industrial multiaxis CO_2 laser systems. For example, in replacing conventional machining, laser cutting has increased throughput in trimming of a gas turbine engine component from 18 pieces per day to 18 pieces in 30 minutes. For another, laser cutting has replaced hand methods for cutting intersecting tubing used in jet aircraft engine ducting. With the hand method, the time to produce one assembly was 1½ hours. With laser cutting, this has been reduced to one minute.

5.2 NUCLEAR AND POWER INDUSTRIES

5.2 (1) Assessing the laser for power plant welding

Clark, J. N., Megaw, J. H. C. and Willgoss, R. A. *Weld. Met. Fabr.*, 47(2): 117-127, 1979.

In a feasibility study the use of laser welding of 6mm. plate in type 316, 310 stainless and Ducol W30 steels was assessed for plant components of 2-10mm. wall thickness. Details of the equipment, welding procedures and mechanical and metallurgical properties of the welds are given and a comparison of laser, electron beam, TIG and plasma welding is made. High quality welds of good reproducibility were obtained by laser welding in the tests. Some solidification cracking was experienced, however, with type 310 stainless steel. Applications are envisaged for laser welding when deep penetration is an advantage and where access is difficult or dangerous, and some possible uses are outlined. (12 refs.)

5.2 (2) Laser welding of steels for power plant

Willgoss, R. A., Megaw, J. H. C. and Clark, J. N. (CEGB, UK). *Opt. Laser Technol.*, 11(12): 73-81, 1979.

Many welding applications in CEGB power plant involve components with wall thicknesses in the range 2-10mm. Therefore, an investigation into the feasibility of laser welding types 316, 310 and Ducol W30 power plant steels in 6mm. thick plate has been made. The welds were made using a high-power, transverse flow, CO_2 laser. The results are compared with TIG, plasma and electron beam welding using capital cost, power requirement, ease of handling and distortion in workpieces as the criteria.

5.2 (3) Development of joining techniques for fabrication of fuel rod simulators

Moorhead, A. J. et al. (Oak Ridge National Laboratory, USA). Rept. No. ORNL-5673, 45 pp., October 1980. Oak Ridge National Laboratory, TN, USA.

Laser welding and furnace brazing techniques are described for joining subassemblies for fuel rod simulators (FRS), that have survived up to 1000h steady-state operation at 700 to 1100°C cladding temperatures and over 5000 thermal transients. A pulsed laser welding procedure that includes use of small diameter filler wire is used to join one end of a resistance heating element to a tubular conductor of an appropriate intermediate material. The other end of the

heating element is laser welded to an end plug, which in turn is welded to a central conductor rod.

5.2 (4) External attachment of titanium sheathed thermocouples to zirconium nuclear fuel rods for the LOFT reactor

Welty, R. K. (Exxon Nuclear Co, Inc, USA). Rept. No. CONF-800719-4, 8pp., 1980. Exxon Nuclear Co, Inc, Richland, WA, USA. Presented at Int. Symp. of Society of Photo-Optical Instrumentation Engineers, 28 July 1980, San Diego, CA.

A laser beam was selected as the welding process because of the extremely high energy input that can be achieved, allowing local fusion of a small area irrespective of the difference in material thickness to be joined. A commercial pulsed laser and energy control system was installed along with specialised welding fixtures. Performance qualifications, and detailed welding procedures were also developed.

5.2 (5) Internal attachment of laser beam welded stainless steel sheathed thermocouples into stainless steel upper end caps in nuclear fuel rods for the LOFT reactor

Welty, R. K. and Reid, R. D. (Exxon Nuclear Co, Inc, USA). Rept. No. CONF-800719-5, 15pp., 1980. Exxon Nuclear Co, Inc, Richland, WA, USA. Presented at Int. Symp. of Society of Photo-Optical Instrumentation Engineers, 28 July 1980, San Diego, CA.

A laser beam was selected because of the extremely high energy input that can be achieved, allowing local fusion of a small area irrespective of the difference in material thickness to be joined. A special weld fixture was designed and fabricated to hold the end cap and the thermocouple with angular and rotational adjustment under the laser beam. A commercial pulsed laser was used to make the welds.

5.2 (6) Pressurisation of nuclear fuel rods using laser welding

King, P. P. (Exxon Nuclear Co, Inc, USA). In, *SPIE Advances in Laser Engineering and Applications*, 31 July- 1 August 1980, San Diego, CA, pp. 24-29. Society of Optical Instrumentation Engineers, USA, 1980.

Pulsed laser drilling and welding of nuclear reactor fuel rods has been demonstrated to be a reliable process for injecting helium into fuel rods at specified pressures and providing an absolute seal. Fuel rods can be pressurised to high pressures.

5.2 (7) Brazing of sensors for high-temperature steam instrumentation systems

Moorhead, A. J. et al. (Oak Ridge Natl. Laboratory, USA). *Weld. J.*, 60(4): 17-28, 1981.

Includes demonstration of procedures developed for brazing a ceramic-to-metal seal and for laser welding of sensor subassemblies into tube walls. Study of three-dimensional phenomena in the upper plenum and core of a pressurised water reactor during the reflood stage of a loss-of-coolant accident.

5.2 (8) Techniques for laser processing, assay, and examination of spent fuel

Gray, J. H., Mitchell, R. C. and Rogell, M. L. (Allied-General Nuclear Services, USA). Rept. No. AGNS-35900-1.2-156, 96pp., November 1981. Allied-General Nuclear Services, Barnwell, SC, USA.

5.2 (9) Application of laser welding to electrical steel strip

Kawasaki Steel Giho, 14(2): 173-181, 1982. (In Japanese)

The application of the CO_2 laser beam to the square butt welding of electrical, Si-steel strip has greatly enhanced weld quality, and resulted in the improved appearance, dimensional accuracy and mechanical properties of the weld. The laser beam is sharp, and precise registering of both the sheared edges of the strip and perfect alignment is required.

5.2 (10) Joining techniques for core flow test loop fuel rod simulators

Moorhead, A. J. and McCulloch, R. W. (Oak Ridge Natl Laboratory, USA). *Nuclear Technol.*, 56(1): 7-22, 1982.

Laser welding techniques have been developed to join subassemblies for fuel rod simulators that have survived up to 1000h steady-state operation at 700 to 1100°C cladding temperatures and over 5000 thermal transients. A pulsed laser welding procedure uses small diameter filler wire to join one end of a resistance heating element to a tubular conductor. The other end of the heating element is laser welded to an end plug, which in turn is welded to a central conductor. Before these welding operations, the intermediate material conductors are vacuum brazed to matching copper leads. The thin walls and ductility of the copper and nickel tubular conductors caused joint machining and fit-up problems.

5.2 (11) Laser fabrication of a model heat exchanger using laser cutting and welding techniques

Ream, S. L. and Veverka, D. B. In, *ICALEO '82, Proc. Materials Processing Symp.*, Vol. 31, 20-23 September 1982, Boston, MA, pp. 141-148. Laser Institute of America, Toledo, OH, USA, 1982.

Laser welding may be applied instead of brazing without the need for a second material and distortion will be minimised. Just as important, the elimination of braze alloy and residual brazing flux in the heat exchanger may provide an improvement in corrosion resistance and a reduction in contamination of high-purity working fluids or gases.

5.2 (12) Use of lasers at the Los Alamos Hot Cell facility

Lazarus, M. E. (Los Alamos National Laboratory, USA). Rept. No. LA-UR-83-804 CONF-830425-7, 8pp. 1983. Los Alamos National Laboratory, NM, USA. Presented to Society of Photo-Optical Instrumentation Engineers Conf., 11 April 1983, Santa Fe, NM, USA.

An optical profilometer that uses a Techmet LaserMike scanning, focused, laser beam, optical micrometer is installed in a remote alpha-gamma containment cell at the Los Alamos Hot-Cell Facility. The Hot-Cell Facility also uses a Korad 20J output ruby pulsed laser to drill a hole in reactor fuel element cladding to sample fission gas. The laser is then used to reweld the hole so that the fuel element will not be contaminated and may be stored without an alpha-containment barrier.

5.2 (13) Welding and brazing of film probe sensor assemblies

Moorhead, A. J. (Oak Ridge National Laboratory, USA). *Weld. J.*, 62(10): 17-27, 1983. Full report available as Rept. No. ORNL-5895, 40pp., November 1982. Oak Ridge National Laboratory, Oak Ridge, TN, USA.

Describes pulsed laser welding and brazing techniques for fabrication of complex sensors for measurement of film thickness and velocity in simulated pressurised-water nuclear reactors. The capability of these sensors to survive repeated exposure to high-temperature steam and water and severe thermal variations was made possible by a ceramic-to-metal seal system based on a metal-dispersion-enhanced ceramic insulator and an experimental brazing filler metal.

The laser subsystem (See 5.2 (15))

5.2 (14) Cutting technology for decommissioning operations. An experimental device for determining the characteristics of and filtering the aerosols produced during plasma-arc cutting of metals

Caropreso, Q. et al. (Comitato Nazionale per la Ricerca e per Lo Svilluppo, d' ell' Energia Nucleare e delle Energie Alternative) CNRS/ENEA, Rome, Italy, 1985, 50pp., (In Italian)

5.2 (15) A laser cutting system for nuclear fuel disassembly

Weil, B. S. (Oak Ridge National Laboratory, USA). In, *Laser Welding, Machining and Materials Processing – Proc. Int. Conf. on Applications of Laser and Electro-Optics, ICALEO '85*, 11-14 November 1985, San Francisco, CA, (Ed. C. Albright), pp. 145–152. Laser Institute of America, Toledo, OH, USA/IFS (Publications) Ltd, Bedford, UK, 1986.

A significant advancement in fuel reprocessing technology has been made by utilising a multikilowatt CO_2 laser to perform cutting operations necessary to remove unprocessible hardware from reactor fuel assemblies.

5.2 (16) Laser drilling of ceramic for heat exchanger applications

Frye, R. W. and Polk, D. H. (United Technologies Research Center, USA). In, *Laser Welding, Machining and Materials Processing – Proc. Int.*

Conf. on Applications of Lasers and Electro-Optics, ICALEO '85, 11-14 November 1985, San Francisco, CA, (Ed. C. Albright), pp. 137-144. Laser Institute of America, Toledo, OH, USA/IFS (Publications) Ltd, Bedford, UK, 1986.

Porous ceramic tiles have a potential application as distributor plates in high temperature fluidised bed waste heat recovery systems. The results of an investigative programme using a Nd:YAG laser for development of processing parameters to install round or slotted holes in silicon carbide, silicon nitride, aluminium oxide or COMPGLAS^{TM} plates will be reported.

5.2 (17) Laser welding of chandelles to the plates of the diagrid employed in the nuclear power plant core

Cai, M. et al. (NIRA, Italy). In, *Proc. 3rd Int. Conf. on Lasers in Manufacturing*, 3-5 June 1986, Paris, France, (Ed. A. Quenzer), pp. 135-144. IFS (Publications) Ltd, Bedford, UK, 1986.

This paper presents the experimental work and the results obtained in laser welding of components, simulating the function of the chandelles,. to the sommier plates of the Super Phoenix 2 (SPX2) nuclear reactor. In the Super Phoenix 1 reactor the fixing of the chandelles to the sommier plates was achieved in a mechanical way by means of threaded ring nuts shut from the inner side.

5.2 (18) Laser welding of electrical sheet steel

Kim, T. K. et al. (Daewoo Heavy Industries Ltd, Korea). In, *Laser Welding, Machining and Materials Processing – Proc. Int. Conf. on Applications of Lasers and Electro-Optics, ICALEO '85*, 11-14 November 1985, San Francisco, CA, (Ed. C. Albright), pp. 59-63. Laser Institute of America, Toledo, OH, WA/IFS (Publications) Ltd, Bedford, UK, 1986.

Using 1kW, single mode CO_2 laser, thickness 0.35mm. grain-oriented electrical sheet steel and thickness 0.5mm. non-oriented sheet steel were butt-welded. Mechanical properties and microstructure were inspected for the weldments. Microhardness traverses were also made from the base sheet through the heat-affected zone (HAZ) and fusion zone. The comparison between laser weldments and TIG weldments was conducted.
From this experiment, it is concluded that laser welding is quite suitable for continuous coil processing in a coil build-up line of the electrical steel industry.

5.2 (19) Nuclear power plant heat exchanger tube cutting by means of a Nd:YAG laser with optical fibre beam transport

Fiorini, O., Germani, G. F. and Pandarese, F. (CISE SpA, Italy). In, *Proc. 3rd Int. Conf. on Lasers in Manufacturing*, 3-5 June 1986, Paris, France, (Ed. A. Quenzer), pp. 19-24. IFS (Publications) Ltd, Bedford, UK, 1986.

Mock-up of the heat exchanger used for tests at CISE; the arrow points the location of the cut (See 5.2 (19))

Describes the apparatus used to cut from the interior a stainless steel tube with inner diameter 17.5mm, 1mm. thick and 9m. from the nearest end. The fibre was put in a stainless steel flexible bellow which assured both the flow of assisting gas, oxygen, and rotation of output coupler around tube axis. A Nd:YAG laser equipped with a quartz 1mm. core diameter, 12m. length fibre optics, was adopted. The input coupler was an 80mm. focal length lens. The output coupler was a single lens imaging system with 1/1 magnification factor. The resulting average power at the

The 5kW CO_2 laser machining system (See 5.2 (20))

working point was 100W with repetitive pulses of 2ms duration and 70Hz repetition rate. The time required for cutting one tube was 66s.

5.2 (20) Some considerations of laser machining system used in an electric company

Mizutame, M. (Toshiba Corporation, Japan). In, *Laser Welding, Machining and Materials Processing – Proc. Int. Conf. on Applications of Lasers and Electro-Optics, ICALEO '85*, 11-14 November 1985, San Francisco, CA, pp. 121-128. Lasers Institute of America, Toledo, OH, USA/IFS (Publications) Ltd, Bedford, UK, 1986.

New manufacturing systems for metal fabrication works in an electric company are presented by inducing laser machining systems that consist of 1kW, 3kW and 5kW laser oscillators. These systems could be combined to prior and next workstations of the line by CAD/CAM systems.

5.3 ELECTRONICS INDUSTRIES

5.3 (1) Laser machining of thermal print heads

Parks, P. F. (P. M. Industries Inc, USA). In, *Int. Microelectron Conf., Proc. of the Technical Program*, 11-13 February 1975, Anaheim, CA and 17-19 June 1975, New York, NY, pp. 146-149. Industrial and Scientific Conference Management Inc, Chicago, IL, USA, 1975.

Machining of thermal print heads by laser is possible on a production basis. The problems encountered can easily be avoided by attention to detail during design and screen preparation. A compromise in the thickness of the layers requiring machining may be desirable to reduce the costs of machining. Higher powered YAG lasers may prove more effective in reducing machining costs.

5.3 (2) Fabrication of an access coupler with single-strand multimode fibre waveguides

Barnoski, M. K. and Friedrich, H. R. (Hughes Research Laboratories, USA). *Appl. Opt.*, 15(11): 2629-2630, 1976.

Describes experiments which show that efficient access couplers can be formed from two 'off-the-shelf' multimode fibres by welding them with a CO_2 laser.

5.3 (3) Laser spot welding and real-time evaluation

Saifi, M. A. et al. (West Electric Co, Inc, USA). *IEEE J. Quantum Electron*, QE-12(2): 129-136, 1976.

The high repetition rate pulsed YAG: laser welding of insulated copper wires to terminal posts and their real-time evaluation with stress wave emission (SWE) are discussed. The experimental and the analytical results show that for the geometry a terminal made out of higher melting-point material than copper is desirable. SWE techniques have been used to predict the quality of a laser weld in real time.

5.3 (4) Laser welding in miniaturised metal-sealed gas protected contact production

Jung, V. and Sauer, M. (Siemens, W. Germany). *Siemens-Z*, 50(4): 265-267, 1976. (In German)

The production of the contact (SGS) developed for switching functions in computerised telephone switching systems such as the EWS electronic switching system requires high precision in order to assure the desired operating reliability and long life. Laser welding has proved to be the most practical and cost-effective method of processing certain parts of the contact.

5.3 (5) Pulse discharge capacitor weight minimisation by peak foil edge fields

Parker, R. D. (Hughes Aircraft Corp, USA). In, *Proc. IEEE Int. Pulsed Power Conf.*, 9-11 November 1976, Lubbock, TX, pp. IIIB2. 1-IIIB2. 6. IEEE (CH1147-8 Reg. 5), New York, 1976.

Corona failure at foil edges is the principal failure mechanism in well-designed and manufactured high energy density pulse discharge capacitors. By forming the foil edge with laser cutting a 25% increase in corona inception voltage over untreated edges is obtained. (4 refs.)

5.3 (6) Thin-film microwave integrated circuits

Aramati, V. S. et al. (Bell Telephone Laboratory, USA). *IEEE Trans. Parts Hybrids Packaging*, PHP-12(4): 309-316, 1976.

Fine-grain alumina and fused silica and low-loss conductor systems that are both solderable and thermocompression bondable were introduced. Further complications which were overcome included bilevel patterns with laser-drilled plated via-holes, the control and measurement of the alumina substrates' dielectric constant, and the use of laser trimming to adjust small geometries of tantalum nitride (Ta//2N) termination resistors. Precise pattern delineation of conductors by laser machining, sputter etching, and selective plating was evaluated and the latter process was found to be the most economical and reliable. (12 refs.)

5.3 (7) Contribution to laser welding of microparts

Trefilova, B. (Welding Research Institute, Czechoslovakia). In, *Laser 77 – Proc. 3rd Opto-Electronics Conf.*, 20-24 June 1977, Munich, W. Germany, pp. 316-321. IPC Science and Technology Press, Guildford, UK, 1977.

At the Welding Research Institute, Bratislava, the equipment VUZ LZ-5 was developed for this purpose; this paper shows its practical employment in welding of microparts made of dissimilar material. (4 refs.)

5.3. (8) Integrated circuit artwork generation by laser machining

Feldman, M. et al. (Bell Laboratory, USA). *Solid State Technol.*, 20(5): 52-57, 1977.

Describes a method for generating reticles used in the step-and-repeat cameras that produce integrated circuit masks. The system utilises a point-by-point raster scan of the entire workpiece. Laser machining is used to write each address. With a YAG laser operating at 300kHz, a 5cm square reticle with a 5μm address structure is machined in less than 30 minutes. The iron oxide laser machined reticles exhibit better linewidth control and higher yield than emulsion reticles.

5.3 (9) Laser-generated resistor capacitor networks

Masopust, O. T. and Saifi, M. A. (West Electric Co, USA). *West Electr. Engng.*, 21(2): 48-58, 1977. Also *IEEE J. Quantum Electron.*, QE-12(2): 120-125, 1976.

A single component containing both resistance and capacitance can be manufactured by using a laser to machine non-conductive lines defining resistance patterns on one or both plates of a metallized-film capacitor. This is made possible by using short laser pulses of much higher intensity than the machining threshold intensity of the zinc. In the manufacturing system used, the lengths of the machined lines are controlled by preset switches, while the spacing between the lines and the line widths are controlled by adjustments within the laser beam optics.

5.3 (10) Micro-circuit flatpack sealing by laser welding

Kirshnaswamy, H. N. and Boccelli, V. E. (Re-Entry and Environmental Systems Division of General Electric, USA). *SAMPE Q.*, 8(4): 11-19, 1977.

The flatpack is a metal case containing conformally coated microcircuits on a ceramic substrate, with leads passing through dielectric glass beads in the case wall. The case and its cover, both made of Kovar, are welded together to form a hermetic seal. Laser welding was developed as a more versatile alternative method to resistance seam welding.

5.3 (11) Optical disc data recorder

Kenney, G. et al. (Philips Laboratory, USA). In, *Digest Papers IEEE 14th Computing Society Int. Conf.*, 28 February – 3 March 1977, San Francisco, CA, pp. 31-32. IEEE, New York, NY, USA.

Recording is done by laser machining of micro-sized pits in a thin tellurium film allowing direct-reading-after-writing of the information.

5.3 (12) The sealing of small-size packages of integrated circuits by laser welding

Ostretsov, Yu. N. et al. *Weld. Prod.*, 24(11): 12-13, 1977.

5.3 (13) Laser welding Cu-Cu and Cu-Al conductors

Jones, M. G. (General Electric, USA). In, *Proc. Technical Program, Electronic Optical Laser Conf. and Exposition*, 19-21 September 1978, Boston, MA, pp. 91-101. Industrial and Scientific Conference Management, Chicago, IL, USA. 1978.

Aluminium and copper joints were welded with a Nd:glass laser operated in a pulse mode at 40-48J per pulse. Strength, electrical resistance and microprobe tests indicate good quality joints. It is shown that, when operating parameters are established for welding ETP copper with a source of monochromatic radiation supplied by a Nd:glass laser, good tensile strength and low electrical resistance can be obtained. With the addition of moderate contact pressure, copper-aluminium conductors can be laser welded resulting in good tensile and impact strengths as well as good electrical resistance stability after thermal cycling.

5.3 (14) Laser welding of electrical interconnections

Bauer, F. r. (Bendix Corp, USA). Rept. No. BDX-613-2022 (Rev.), 26pp., December 1978. Bendix Corp, Kansas City, MO, USA.

Processes and equipment have been developed for welding thin aluminium and copper foils using a Nd:YAG laser. Laser welding provides an alternative technique with improved quality for welding these types of electrical connections.

5.3 (15) Production of microwave integrated circuits by laser machining

Tomlinson, J. (EMI Electronics, UK). *Radio Electron. Engng.*, 48 (1-2): 43-46, 1978.

Discusses laser machining as an alternative to photolithography for the production of microwave integrated circuits. A complete system is described comprising a laser with computer-controlled XY table moving a prepared substrate through the laser beam. The procedure for *transfering the design sketch of a microwave circuit into a finished substrate is described.*

5.3 (16) The pulsed laser welding of conductors to films in the manufacture of micro-devices

Avramchenko, P. F., Velichko, O. A. and Moravskii, V. E. (Academy of Science USSR). *Autom. Weld.*, 31(5): 20-21, 1978.

It has been proved that copper, gold and aluminium wires can be joined to thin metallic films on Sital by pulsed laser welding; the diameter of the wire is 40-50nm. The welded joints have satisfactory strength, and there are no defects in the welded spot or the Sital alloy.

5.3 (17) YAG laser cutting of c-axis sapphire

Kubo, Larry Y. (Hewlett-Packard Co, USA). In, *Proc. 23rd Electronic Components Conf.*, 24-26 April 1978, Anaheim, CA, pp. 85-86. IEEE, New York, NY, USA, 1978.

Thin film networks of conductor traces, resistors, and capacitors are fabricated on c-axis sapphire, the electrical characteristics of which are advantageous for circuitry in the microwave range. C-axis sapphire requires complete cutting or sawing because its crystalline structure causes scribed substrates to cleave at oblique angles. Diamond blade sawing has been found to produce mechanically induced chips and fractures in unacceptable degree and size on 10mm. thick c-axis sapphire. Laser cutting overcame these mechanically induced problems.

5.3 (18) Interference effects in laser micromachining of thin films on silicon

Kestenbaum, A. (West Electric Co, USA). *J. Appl. Phys.*, 50(7): 5012-5017, 1979.

Examines the laser machining process at 1.06µm for a tantalum nitride thin-film resistor material on top of a dielectric film on silicon. The existence of strong optical interference effects in laser machining of these resistors is verified, and their influence on various laser-trimming parameters is determined.

5.3 (19) Laser cutting metal through quartz

Aggarwal, B. K. and Gandy, F. D. (IBM Corp, USA). *IBM Tech. Disclosure Bulletin*, 22(5): 1971-1972, 1979.

A semiconductor substrate carries an Al conductor which is to be cut by means of the laser beam after it has been completed with two layers of quartz. This improvement consists in

shaping the upper quartz layer so as to have minimum thickness over the area of the conductor to be cut.

5.3 (20) Laser-welded package for semiconductor and superconductor electronics

Melcher, R. L. (IBM Corp, USA). *IBM Tech. Disclosure Bulletin*, 22(3): 834, 1979.

A method for laser welding of semiconductor chips to a substrate is described.

5.3 (21) Laser welding of exhaust gas oxygen sensor

Chang, U. I. and Casey, K. W. (Ford Motor Co, USA). In, *Proc. Applications of Lasers in Material Processing Conf.*, 18-20 April 1979, Washington, DC, pp. 51-64. ASM (Materials/Metalwork Technology Series), 1979.

In joining of Ni-Cr lead wires to terminal pins of an exhaust gas oxygen sensor, functional laser welds can be made with a wide variation of welding parameters. The Ni-Cr lead wire was used as the filler metal and a defocused laser beam as the heat source. (7 refs.)

5.3 (22) Laser welding of precision instruments

Gensmer, W. (Honeywell Avionics, USA). In, *Proc. Technical Program, Optical Laser Conf. and Exposition*, 23-25 October 1979, Anaheim, CA, pp. 257-260. Industrial and Scientific Conference Management Inc, Chicago, IL, USA, 1979.

Examines the application of lasers as heat sources for the welded fabrication of precision instruments in low volume assembly units. The merits of laser weld fabrication designs are compared with conventional welding for cost effectiveness.

5.3 (23) Laser welding speeds relay production

Insul. Circuits, 25(4): 24-25, 1979.

Describes a recently developed computer controlled laser system which spot welds terminals on miniature relays four times faster than the resistance welding system it replaced. Operating at speeds up to 20 welds per second, the new system offers reliability, speedier inspection, and increased production rate. Use on printed circuit boards in telephone transmission equipment is briefly described.

5.3 (24) Laser welding technology for heat resistant wire tensoresistors of alloy kH20N80

Kokhanovskii, V. D. et al. *Svar. Proizvod.*, (9): 13-14, 1979. (In Russian)

The technology was developed for welding the exit 0-1mm. diameter conductors to the grid of heat-resistant wire tenso-resistors 0.03mm. diameter of alloy kH20N80. In comparison with other forms of welding of the grid, the laser welding increased the static and fatigue strengths of welded components.

5.3 (25) Reliability of LSI memory circuits exposed to laser cutting

Rand, M. J. (Bell Laboratory, USA). in, *Proc. 17th Annual Reliability Physics Symp.*, 24-26 April 1979, San Francisco, CA, pp. 220-225. IEEE, New York, NY, 1979.

Laser activation of spare row or column elements of VLSI memories is a powerful tool for yield enhancement, and is presently in use on a 64k-bit MOS RAM. Experiments were carried out on the effects of the laser on materials and structures of the device. Also, laser-cut 16k RAM testers and polysilicon resistors were bias-temperature aged, with note made of electrical drifts which might indicate penetration of contaminants at the laser-cut site. (11 refs.)

5.3 (26) Die board cutting by CO_2 laser

Forbes, N. (Ferranti Ltd, UK). In, *Proc. SPIE Advances in Laser Engineering and Applications Conf.*, 31 July – 1 August 1980, pp. 8-17. SPIE, Bellingham, WA, USA, 1980.

The conventional method of dieboard manufacture is described and compared with the laser dieboard process. Consideration is given to savings achievable as a result of the precision die produced by laser cutting in combination with numerical control.

5.3 (27) Effects of xenon cover gas in CO_2 laser welding

Hendrix, T. L. (Bendix Corp, USA). Rept. No. BDX-613-2442, 31pp., July 1980. Bendix Corp, Kansas City, MO, USA.

Argues that weld spatter in CO_2 laser welding is detrimental to miniature components and discusses the effects of using xenon gas as an inert laser welding atmosphere to reduce weld spatter. The laser plume characteristics, weld penetration, and weld spatter are evaluated.

5.3 (28) Laser cold processing takes the heat off semiconductors

Kaplan, R. A., Cohe, M. G. and Kiu, K. C. (Quantronix Corp, USA). *Electronics*, 53(5): 137-142, 1980.

5.3 (29) Packaging and production

Lyman, J. *Electronics*, 53(23): 222-226, 1980. See also Ibid 52(22): 216-218, 1979.

Discusses the development of the use of the laser in welding and chip and board processing. The use of the laser for annealing silicon wafers damaged by ion implantation, writing patterns on resist covered printed circuit boards, and welding flexible circuits to connectors is examined.

5.3 (30) Dual lasers speed termination of flexible printed wiring

Henderson, J. A. (Westinghouse Electric Corp, USA). *Electronics*, 54(19): 149-154, 1981.

Describes a new approach in linking flexible printed wiring to connectors that is commercially as well as militarily viable. The new approach is capable of being used with any planar connector having one or two rows of conductor pins on its back. The cables are stripped by a CO_2 laser beam, and only half of the wire's insulation need be removed. Welding flexible printed wiring to connectors using a Nd:YAG laser yields a highly reliable joint.

5.3 (31) Fabrication of foil masks using laser cutting

De Silva, G. M. S., Leather, J. A. and Anderson, J. C. (Imperial College, UK). *Thin Solid Films*, 77(4): 341-346, 1981

Describes a method of fabricating foil masks with micron dimensions using a Nd:YAG laser. The minimum cut width obtainable with the system is 2.5µm and foils up to 100µm thick could be cut. The cutting parameters and the mask patterns which can be obtained with the system are explained.

5.3 (32) Flexible printed circuits with integral molded connectors. A manufacturing methods and technology program

Henderson, J. A. (Westinghouse Defense and Electronic Systems Center, USA). Final Rept. No. AD-A098 02413, 148pp., March 1981. Westinghouse Defense and Electronic Systems Center, Baltimore, MD, USA. See also rept. no. AD-A098 025/0.

Through the use of industrial laser technology, new high speed epoxy developments, and microprocessor controlled automation, processes for termination of flexible printed wiring (FPW) to connectors have been developed which can result in 6 to 1 cost reduction of terminated systems with significantly improved system reliability and maintainability.

5.3 (33) Flexible printed circuits with integral molded connectors. Automated facilities report

Hall, R. L. (Westinghouse Defense and Electronic Systems Center, USA). Rept. No. AD-A098 025/0, 50pp., March 1981. See also rept no. AD-A098 024/3. Westinghouse Defense and Electronic Systems Center, Baltimore, MD, USA.

Three new processes used in termination of flexible printed wiring (FPW) to connectors were developed on this programme. They are laser ablation of insulation by CO_2 laser, laser welding by Nd:YAG laser, and liquid injection moulding of small parts. The integration of these processes into a fully automated facility capable of one assembly per minute production was then projected.

5.3 (34) Laser-patterned Ta_2N resistors for thin-film circuits

Scarff, P. L. and Kiszka, L. J. (Bell Laboratory, USA). In, *Proc. 31st Electronic Components Conf.*, 11-13 May 1981, Atlanta, GA, pp. 449-455. IEEE, Piscataway, NJ, USA, 1981. Also *IEEE Trans. Components Hybrids Manufacturing Technololgy*, CHMT-4: 361-366, 1981.

A process using laser machining is developed for patterning the detailed features of thin-film Ta resistors on ceramic substrates. Laser machining serpentine patterns in etched blocks of resistor film reduces line and space widths below those practical with off-contact photolithography and wet etching.

5.3 (35) Laser tool pulsed solid state laser for welding and ablation of material

Seiler, P. (Bundesministerium fuer Forschung und Technologie, W. Germany). Rept. No. BMFT-FB-T-81-043, 20pp., February 1981. (In German) BMFT, Bonn-Bad Godesberg W. Germany. (Translation p 1382-238247, 14 pp., July 1982)

The incorporated Nd:glass pulsed lasers have pulse energies up to 100Ws and average output powers up to 100W. The machines are used in precision and electrical engineering, mainly combined with industrial handling. This technique has a wide range of applications in the field of spot and seam welding of small parts and is widely used in manufacturing contacts in the field of electronics.

5.3 (36) Laser welding in microtechnology

Wuthrich, R. *Rev. Polytech.*, (4): 445-447, 1981. (In French) See also *Bulletin Ann. Societe Suisse de Chorometrie et Laboratoire Suisse de Recherches Horlogeres, 55th Congress of Swiss Chorometry Society*, Vol. 9, 10-11 October 1980, Berne, Switzerland, pp. 159-164.

Besides welding, the technique can be used, amongst other things, for cutting and drilling. Its advantages, which include no contact with the workpiece, no tool wear, and highly accurate performance, are listed. Usually solid (Nd:YAG and CO_2 lasers are used. The solid laser is used for high-power tasks, e.g. hardening steel, drilling surgical needles, adjusting hybrid circuit component values, and the gas laser for such tasks as the adjustment of the frequency of quartz oscillators. The optical system for practical applications and fixed beam, beam scanning, beam splitting and time sharing techniques is examined. The control of the laser output so that the energy density is constant at the target is also considered.

5.3 (37) Novel concepts in laser welding of miniature relays

Califano, V. E. and Muniz, J. M. (West Electric, USA). In, *Proc. 29th Relay Conf*, 28-29 April 1981, Stillwater, OK, pp. 10.1-10.4. National Association of Relay Manufacturers, Elkhart, IN, USA, 1981.

Describes steps leading to the introduction of laser welding methods of miniature relays. Systems concepts and system architecture are dealt with.

5.3 (38) Investigations of holes machined by laser beam on Al and Cr thin films

Yamada, K. et al. (Mitsubishi Electric Corp, Japan). *J. Appl. Phys.*, 53(4): 3231-3236, 1982.

Transmission and scanning elcctron microscopy investigations of holes created by laser irradiation on Al and Cr films were carried out. The results showed that both films have the well-shaped round holes and that the peripheral regions form gentle swells for Al films and steep swells for Cr films. From the measurement of surface shape across the groove made by a laser beam, the height of swells was nearly the same as the film thickness. It was also found that the swells are composed of large recrystallised grains, and that the effect of the existence of large grains on the recording characteristics is not so evident. A numerical analysis of temperature rise of the metal films was made.

5.3 (39) Laser annealing of semiconductors

Poate, J. M. and Mayer, J. W. (Eds.) Academic Press, 1982, 592pp.

5.3 (40) Laser butt welding of copper and gold wire

Salzer, T. E. In, *ICALEO '82, Proc. Materials Processing Symp.*, Vol. 31, 20-23 September 1982, Boston, MA, pp. 73-78. Laser Institute of America, Toledo, OH, USA, 1982.

Several welding techniques were considered, among which were percussive welding, plasma arc welding and laser welding. Initial feasibility experiments indicated that, while each of these welding techniques showed promise, laser welding results were so outstanding that the balance of the study concentrated on laser welding exclusively.

5.3 (41) Laser lead welding

Jones, M. G. In, *ICALEO '82, Proc. Materials Processing Symp.*, Vol. 31, 20-23 September 1982, Boston, MA, pp. 87-100. Laser Institute of America, Toledo, OH, USA, 1982.

Pulse laser systems which emit energy in the near infrared or visible frequency spectrum are ideal for welding small electrical conductor leads of similar and dissimilar materials (Cu and Al) which may also possess different sizes. Laser welding of leads is desirable when low heat input is required, insulation degradation is of concern and potential automation is being considered. Welding by laser will result in good electrical and mechanical joint properties.

5.3 (42) Laser welding of microelectronic components

Dane, K. *Schweisstechnik*, 32(3): 112-114, 1982. (In German)

The laser welding process was used to weld on 0.4nm diameter tungsten or beryllium bronze wire to the outer edge of an 0.4mm. thick Nicosil sheet. A Nd:YAG laser gives satisfactory strength and low thermal distortion when used to weld high precision microelectronic components.

5.3 (43) Mass-production technology for small electron guns of color picture tubes mainly based on laser welding

Toshiba Corp, Japan. *Res. Dev. Jpn.*, pp. 65-71, 1982. Awarded Okochi Memorial Prize.

The performance of such elements as picture quality, reliability and power consumption depend largely on the characteristics of the electron gun as the most important components of colour picture tubes. The desire to reduce power consumption and save resources is increasing demands for colour picture tubes with a smaller neck diameter, which requires electron guns to be made smaller. Since this reduction in size results in a smaller lens system in the electron gun, it is necessary to overcome severe restrictions on design while aiming at higher accuracy of parts and assembly in order to ensure adequate picture quality and reliability. To meet these requirements high-accuracy and high-quality assembly techniques based on laser welding have been developed.

5.3 (44) Micro-materials processing by laser beam

Cohen, M. G., Kaplan, R. A. and Arthurs, E. G. (Quantronix Corp, USA). *IEEE Proc.*, 70(6): 545-555, 1982.

State-of-the-art report with 34 references.

5.3 (45) MLC laser sizing and cutting using retrace

Chiaiese, V. C., et al. (IBM Corp, USA). *IBM Tech. Disclosure Bulletin*, 25(38): 1492-1493, 1982.

Through the use of retrace, laser cut edge quality on MLC (multilayer ceramic) green laminates is improved. (One of several papers in this issue on laser sizing systems: see also pp. 1465-1468 – alignment and distortion correction; pp. 1469-1470 – laser sizing system vacuum chuck; pp. 1471-1472 – scrap removal system; pp. 1473-1475 – laminate handling; pp. 1442-1443 – edge characteristics; pp. 1490-1491 – gas mix optimisation; pp. 1492-1493 – cooling system.)

5.3 (46) A demonstration of very large area integration using laser restructuring

Raffel, J. I. et al. (Lincoln Laboratory, MIT, USA). In, *Proc. 1983 IEEE Int. Symp. on Circuits and Systems*, Vol. 2, 2-4 May 1983, Newport Beach, CA, pp. 781-784. IEEE, New York, NY, USA, 1983.

5.3 (47) High speed laser machining of small diameter holes in printed circuit substrates

Zhuk, F. I. et al. *Phys. & Chem. Mater. Treat.*, 17(3), 5pp., 1983.

Pulses of microsecond duration were used, and the power of the pulse was around 1kW in the wavelength range 10.6nm. It was established for the corresponding focusing system at laser pulse repetition frequencies of up to 1.5kHz that there are optimum pulse sequences at which the conicity of the holes is minimal and the material at the edges of the holes does not oxidise.

5.3 (48) Resistance seam and laser welding of large hybrid metal packages

Stockham, N. R. and Dawes, C. J. (Welding Institute, UK). In, *Int. J. Hybrid Microelectronics*, 6(1): 509-519, 1983. (Int. Microelectronics Symp., 1983, 31 October- 2 November 1983, Philadelphia, PA.)

Includes an investigation into laser welding of Fe-Ni-Co alloy packages, plated with Au or electroless or electrolytic Ni.

5.3 (49) Laser adjustment of linear monolithic circuits

Litwin, A. (RIFA Sweden), and Smart, D. V. (Teradyne, Inc, USA). In, *ICALEO '83, Proc. Materials Processing Symp.*, Vol. 38, 14-17 November 1983, Los Angeles, CA, (Ed. E. A. Metzbower), pp. 166-173. Laser Institute of America, Toledo, OH, USA, 1984.

Laser trimming can be performed on conventionally processed linear circuits to improve both yield and performance. A study was made comparing aluminium links and polysilicon fuses as well as trimming polysilicon resistors. Material performance over a range of link widths, passivation layer thicknesses, laser energies and laser pulse width were investigated.

5.3 (50) Laser processing of semiconductors: an overview

Wood, R. F., White, C. W. and Young, R. T. In, *Semiconductors and Semimetals, Vol. 23: Pulsed Laser Processing of Semiconductors*, pp. 1-41. Academic Press, Orlando, FL, USA, 1984.

Provides a combined state-of-the-art and historical survey of the development of laser processing of semiconductors.

5.3 (51) Laser reflow soldering of film capacitors

Fanning, W. J. (Western Electric Co, USA). In, *ICALEO '83, Proc. Materials Processing Symp.*, Vol. 38, 14-17 November 1983, Los Angeles, CA, (Ed. E. A. Metzbower), pp. 180-191. Laser Institute of America, Toledo, OH, USA, 1984.

Manufacture of miniature film capacitor now utilises three laser-based processes. Paper highlights a CO_2 laser used to reflow solder 30,000 capacitor elements to terminal frames each hour of operation. Also described is a YAG laser which demetallises capacitors' aluminised polyester films during winding and a CO_2 laser marking system.

5.3 (52) Laser spot welding in the electronics industry

Notenboom, G. J. A. M. (Welding Institute, UK). In, *Laser Welding, Cutting and Surface Treatment*, pp. 36-39. Welding Institute, Cambridge, UK, 1984.

Concentrates on spot welding with Nd:YAG or Nd:glass laser systems. Advantages over resistance welding are discussed in relation to weld geometry and metallurgical behaviour. Beam handling is used to reduce investment costs, and in addition leads to better accessibility for the welding machine. The assembly of electron guns for TV tubes is also discussed, as a typical application of Nd:YAG laser welding.

5.3 (53) Laser-welding the large MIC: a new approach to hermetic sealing

Simpson, G. (M/A-COM Inc, USA). *Microwave J.*, 27(11): 169-179, 1984.

Discusses the application of laser welding technology to the problem of generating true fusion seals. Laser technology is making a major impact on the world of MIC sealing. The technology makes it possible to provide deep, dense, reliable hermetic welds in high-conductivity materials, especially aluminium, and is clean.

5.3 (54) Laser welding in the manufacture of heart pacemakers

Janssen, G. W. G. (Medtronic BV, Netherlands). In, *Laser Welding, Cutting and Surface Treatment*, pp. 33-35. Welding Institute, Cambridge, UK, 1984.

Metal housings for the internal electronics and power source have now become standard for all pacemaker manufacturers. Further, welding the metal parts is essential to ensure the hermetic sealing of every device. This paper proposes laser welding as a key process in the manufacture of pacemakers.

5.3 (55) Low cost hermetic sealing of microwave modules

Mendicino, P. A. (Westinghouse Electric Corp, USA). In, *Proc. 1984 Int. Symp. on Microelectronics*, 17-19 September 1984, Dallas, TX, pp. 213-217. International Society of Hybrid Microelectronics. Montgomery, AL, USA, 1984.

5.3 (56) Machining multilayer circuit boards – Part 2

Hagge, J. K. and Mather, J. C. (Rockwell International, USA). *Circuit World*, 11(1): 18-29, 1984.

Ultrasonic, chemical and laser machining methods are briefly outlined along with drill geometry, drilling equipment and operating parameters.

5.3 (57) Automation of production process for current conducting coatings in printed circuit board openings by a laser-electrical method

Suminov, I. V. *Mekh. & Avtom. Proizvod.*, (10): 2-3, 1985. (In Russian)

A solid state laser in free generating mode is used with a pulsed plasma accelerator based on planar electrodes and a dielectric insulator. Tests indicate that metallisation can be successfully carried out for hole sizes down to around 0.1mm and depth up to 4mm giving coating thicknesses in the range 5-15 microns.

5.3 (58) Hybrid microelectronic laser machining requirements

Parks, P. F. (P/M Industries Inc, USA). In, *Proc. SPIE Applications of High-Power Lasers*, Vol. 527, 22-23, January 1985, Los Angeles, CA, pp. 51-55. International Society of Optical Engineers, USA.

Briefly describes resistor trimming and ceramic substrate machining, drawing attention to unresolved problems in the economic use of lasers in this area.

5.3 (59) Experience with laser on automated spotwelding duties

Seiler, P. *IPE Int. Ind. Prod. Engng.*, 9(2): 74-77, 1985.

A production method in which solid-state lasers, i.e. Nd:glass or Nd:YAG lasers, have been used for the welding of electron guns for the picture tubes of television receivers, VDUs for data processing equipment and other applications is reported. Up to 50 spot welds are required/system. (5 refs.)

5.3 (60) Hermetic sealing of Kovar hybrid packages by laser welding

Norrman, S. (Ericsson Radio Systems, Division of Microelectronics, Sweden). *Hybrid circuits*, (8): 21-23, 1985.

Laser welding of Kovar (54% Fe, 16% Co, 29% Ni, <1% Mn, Cr) has been evaluated in order to develop a reliable welding method for hermetic encapsulation of hybrid packages. In particular, the design of the welding joint and the impact of Au/Ni plating on the joint quality have been examined.

5.3 (61) Laser annealing of semiconductors

Bertolotti, M. (Rome University, Italy). *J. Sov. Laser Res.*, 6(4): 395-404, 1981. (Use of Lasers in Atomic, Molecular and Nuclear Physics – Proc. 2nd AU-Union School, 29 June – 7 July 1981, Vilnius, USSR.

The paper discusses: mechanism of heating; main characteristics of laser annealing of semiconductors; and low power effects.

5.3 (62) Laser cutting of aluminium stripes for debugging integrated circuits

Yamaguchi, H. et al. (Hitachi Ltd, Japan). *IEEE J. Solid-State Circuits*, SC-20(6): 1259-1264, 1985.

5.3 (63) Laser technology in printed assembly plate production

Machulka, G. A. et al. *Telecommun. & Radio Eng. Part 2*, 40(3): 50-56, 1985.

Results are reported of investigations on the use of laser radiation in manufacturing printed assembly plates and the duration of the technological cycle and productivity (the application of protective patterns, drilling of holes, soldering of non-printed components, etc.).

5.3 (64) Laser trimming works well in some automotive applications

Stickney, B. and Johnson, R. (Electro-Science Industries, Inc, USA). *Electron. Packag. & Prod.*, 25(4): 154-156, 1985.

The laser is used to trim circuit components to achieve certain circuit characteristics. One specific application for a laser trim is the Ford Silicon Capacitance Absolute Pressure sensor (SCAP). The SCAP is a thick-film hybrid air-pressure sensor to be used with Ford's electronic engine control system.

5.3 (65) Present and future applications for laser processing of hybrids and semiconductors

Swenson, E. J. (Electro-Science Industries Inc, USA). In, *Proc. SPIE applications of High-Power Lasers*, Vol. 527, 22-23 January 1985, Los Angeles, CA, pp. 45-50. International Society of Optical Engineers, USA.

Current applications covered include thick-and thin-film resistor trimming, deposited film and polysilicon resistors on silicon trimming and redundant memory repair. Emerging applications include microcircuit mask making and capacitor trimming. Examples of processes still under development include selective annealing. (27 refs.)

5.3 (66) Sub-micron array personalisation using radiation effect

IBM Tech. Discl. Bulletin, 27(10B); 6091-6092, 1985.

The laser welding technique has been used to rapidly personalise a bi-polar programmable logic array. Since FET devices are sensitive to laser-indiced charges, laser welding is not applicable to the personalisation of devices as small as a 2μm array. Therefore a laser microfab technique has been developed. Describes a proposed technique using E-beam or X-ray radiation effect to solve this problem. The radiation effect is based on the radiation damage caused by the E-beam or the X-ray during the direct writing of silicon gate wafers.

5.3 (67) Technological processing of materials by means of radiation from a sweep laser. I. Laser cutting of thin film resistors

Galich, G. A. et al. *Kvantovaya Elektron.*, (28): 31-39, 1985. (In Russian)

Describes a system which employs a sweep laser and a focused laser beam in the cutting of thin-film resistors. A relationship is established linking the rate of removal of the resistive layer, the degree of overlap of the focused spots and the length of the cut with parameters of a sweep laser based on Nd:glass and the angular dispersion of the diffraction grating used for sweeping the laser beam. (18 refs.)

5.3 (68) A wafer-scale digital integrator using restructurable VLSI

Raffel, J. I. et al. (MIT Lincoln Laboratory, USA). *IEEE Trans. Electron Devices*, ED-32(2): 479-486, 1985.

Wafer-scale integration has been demonstrated by fabricating a digital integrator on a monolithic $20cm^2$ silicon chip. Large-area integration is accomplished by laser-programming of metal interconnect for defect avoidance. Describes the technology for laser welding and cutting, the design methodology and CAD tools developed for wafer-scale integration, and the integrator itself. (24 refs.)

5.3 (69) Sensitive mapping of latch-up in CMOS under laser-scan irradiation

Auvert, G. (CNET, France) Chion, A. and Mackowiak, E. (Thompson, France). In, *Laser Welding, Machining and Materials Processing – Proc. Int. Conf. on Applications of Lasers and Electro-Optics, ICALEO '85*, 11-14 November 1985, San Francisco, CA, (Ed. C. Albright), pp. 221-228. Laser Institute of America, Toledo, OH, USA/IFS (Publications) Ltd, Bedford, UK, 1986.

Inducing latch-up with a focused and scanned laser beam is a powerful technique for locating small sensitive areas in a CMOS integrated circuit. This is a simple, effective and non-destructive technique for quantitatively measuring the latch-up sensitivity in VLSI circuits. Using this laser beam technique, the influence of the design rules on the latch-up characteristics of CMOS microprocessors has been established.

5.3 (70) Silicon deposition by photolytic or pyrolytic dissociation of silane under IR laser irradiation and characterisation of films

Tonneau, D., Auvert, G. and Pauleau, Y. (Centre National d'Etudes des Telecommunications, France). In, *Laser Welding, Machining and Materials Processing – Proc. Int. Conf. on Applications of Lasers and Electro-Optics, ICALEO '85*, 11-14 November 1985, San Francisco, CA, (Ed. C. Albright), pp. 213-220. Laser Institute of America, Toledo, OH, USA/IFS (Publications) Ltd, Bedford, UK, 1986.

Laser chemical techniques are investigated for material processing in microelectronic technology. Pulsed and CW CO_2 lasers are used for deposition of amorphous or polycrystalline silicon films from decomposition of silane. The deposition rate and spatial resolution obtained with the CW laser are convenient for direct writing technology.

5.4 TESTING AND INSPECTION

5.4 (1) A laser-based technique for alignment and deflection measurement

Harrison, P. W. (Department of the Environment, UK). *Civil Eng. & Pub. Wks. Rev.*, 68(800): 224-227, 1973.

Describes use of the laser technique in several applications including measurements carried out for the DOE of the form of box girders in a number of bridges and viaducts in England.

5.4 (2) Laser beam profiler and detector-surface scanner

Foulk, L. R. (Bendix Corp, USA). Rept. No. BDX-613-1841 (Rev.), 32pp., November 1978. Bendix Corp, Kansas City, MO, USA.

Describes the development of a portable, relatively inexpensive laser beam profiler and detector-surface scanner to map the intensity profile of laser beams and the cross-sectional response of optical detectors within a wavelength range of 633nm to 10.6µm. Equipment diagrams, circuit schematics, and examples of detector-response maps and laser beam profiles are provided.

5.4 (3) Speckle interferometry and other inspection techniques

Butters, J. N. (Loughborough University of Technology, UK). In, *Proc. 1st Int. Conf. on Lasers in Manufacturing*, 1-3 November 1983, Brighton, UK, (Ed. M. F. Kimmitt), pp. 149-160. IFS (Publications) Ltd, Bedford, UK, 1983.

Interferometry and the associated techniques of projected grid measurement are discussed in relation to industrial measuring tasks – from those requiring the ultimate in resolution, to those where resolution of the order of a millimetre only is required. Whilst the interferometer techniques are stand-alone devices as measuring sensors, they are related in the paper to the growing technology of vision systems applied to industrial measurement.

5.4 (4) Part inspection with a laser measurement system

Drake, E. S. (Hewlett-Packard Ltd, UK). In, *Proc. 1st Int. Conf. on Lasers in Manufacturing*, 1-3 November 1983, Brighton, UK, (Ed. M. F. Kimmitt), pp. 143-148. IFS (Publications) Ltd, Bedford, UK, 1983.

Accurate measurement of close tolerance machined parts is an excellent application for a laser measurement system. This application usually requires some equipment fabrication by the user, and several solutions to this possible problems are discussed. Generally, these solutions involve mechanical means for translating part dimensions to a linear movement of the laser system's optical devices. Through these techniques, the high accuracy and resolution of the laser system can be realised.

A group of helium-neon lasers (See 5.4 (3))

5.4 (5) Lasers: a new tool for engine designers

Dale, B. W. (Harwell Laboratories, UK). In, *Proc. 1st Int. Conf. on Lasers in Manufacturing*, 1-3 November 1983, Brighton, UK, (Ed. M. F. Kimmitt), pp. 109-116. IFS (Publications) Ltd, Bedford, UK, 1983.

Examines laser anemometry, droplet and particle sizing, and Ramon scattering. The applications of these optical techniques are then briefly discussed.

5.4 (6) Application of holographic interferometry to piston deformation

Evans, W. T. and Russell, R. M. (The Polytechnic of Wales, UK). In, *Proc. 1st Int. Conf. on Lasers in Manufacturing*, 1-3 November 1983, Brighton, UK, (Ed. M. F. Kimmitt), pp. 181-192. IFS (Publications) Ltd, Bedford, UK, 1983.

The current study is concerned with measuring the thermal distortion experimentally using holographic interferometry which is well suited for the purpose, being a non-contacting field method. In order to find the absolute displacement at a given point on the piston, three displacement vectors must be constructed at that point and to this end three separate holograms were simultaneously obtained from different viewing positions. Photographs were taken through each of the resulting holograms, the fringe patterns digitised and used as input to a computer program which produced the final displacement patterns.

5.4 (7) Development of a high resolution in-process sensor for surface roughness by laser beam

Mitsuki, K. et al. (Mechanical Engineering Laboratory, Japan). *Bulletin Jpn. Soc. Precis. Engrs*, 19(2): 142-143, 1985.

Recent developments in machining processes, especially in the field of single point diamond turning or semiconductor technology, require the development of a non-contact precise in-process sensor for surface roughness. Applying the focus detection method, a high resolution in-process surface roughness sensor has been developed.

5.4 (8) The use of electronic speckle pattern interferometry (ESPI) as an inspection tool

Montgomery, P. C. and Tyrer, J. (Loughborough University of Technology, UK). In, *Proc. 2nd Int. Conf. on Lasers in Manufacturing*, 26-28 March 1985, Birmingham, UK, (Ed. M. F. Kimmitt), pp. 141-150. IFS (Publications) Ltd, Bedford, UK, 1985.

Electronic Speckle Pattern Interferometry (ESPI) has become a useful laser technique for observing surface deformations of less than 1μm over a component area ranging from 1mm^2 to 1m^2. Results, displayed as a contour map on a TV monitor, provide information helpful to design, quality control and in-service testing. This paper presents typical applications of the technique to component testing in manufacturing. The authors also discuss current and future advances for improving performance and reducing the cost of ESPI systems.

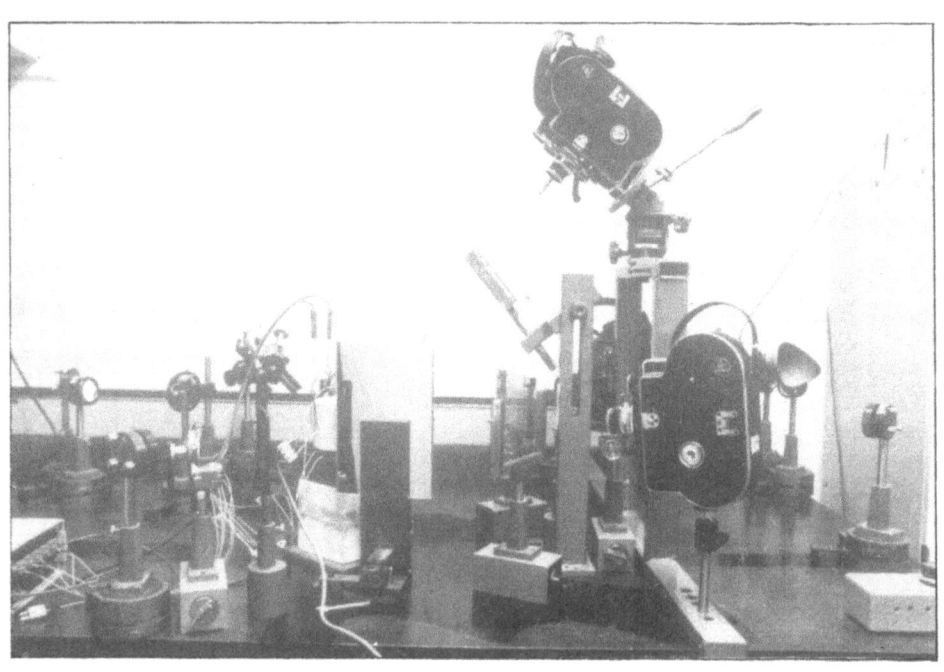

Experimental set-up for holographic interferometry (See 5.4 (6))

Optical bench and electronics for ESPI showing how it can be used in a noisy environment (See 5.4 (8))

5.4 (9) Optical polar profilometer: a new method for analysis of surfaces with circular symmetry

Laufer, G. et al. In, *Proc. 2nd Int. Conf. on Lasers in Manufacturing*, 26-28 March 1985, Birmingham, UK, (Ed. M. F. Kimmitt), pp. 169-172. IFS (Publications) Ltd, Bedford, UK, 1985.

A non-contact optical measuring method is presented. This technique allows the recording polar curves of surfaces with circular symmetry at a high accuracy which is obtained by using differential detection. It was demonstrated that surfaces may be analysed at an accuracy which exceeds 2.5µm.

5.4 (10) Application of immersion technique on the measurement of some engineering products

El Sayed, M. M. (Helwan University, Egypt), and Koura, M. M. (Ain Shams University, Egypt) In, *Proc. 2nd Int. Conf. on Lasers in Manufacturing*, 26–28 March, 1985, Birmingham, UK, (Ed. M. F. Kimmitt) pp. 159–168. IFS (Publications) Ltd, Bedford, UK, 1985.

This paper shows the feasibility of the holographic immersion technique in mapping the contour of some engineering products which presents the inspection personnel in industry with various difficulties. Also includes holographic interferometry patterns which were taken for some objects together with the qualitative and quantitative analyses that illustrate the technique.

AUTHOR INDEX

A
Abe,N.,3.4.(5);3.4.(17)
Aberkane,K.,3.1.(10)
Aberman,Z.,2.2.(14)
Abil'siitov,G.A.,1.4.(35)
Aden,R.J.,3.1.(4)
Aerni,G.,2.1(24)
Aggarwal,B.K.,5.3.(19)
Ahmed,A.U.,3.3.(29)
Akau,R.L.,3.4.(20)
Albert,M.,5.1.(16)
Albright,C.E.,1.4.(54);3.4.(2);3.4.(10);
  4.1.2.2.(29)
Alekseev,V.A.,4.1.2.2.(8)
Alexander,J.,4.1.1.(17);4.2.5.1.(5)
Allen,D.M.,2.3.(4)
Allen,R.,4.3.5.(2)
Allmen,M.V.,4.2.4.2.(1)
Amende,W.,4.2.,5.1.(9)
Anderson,D.G.,1.4.(6);2.1(2)
Anderson,J.C.,4.3.5.(3);5.3.(31)
Anderson,R.B.,4.3.5.(6)
Andrews,J.G.,3.3.(2)
Andriyakhin,V.M.,4.3.2.(5);4.4.2.(4)
Angus,J.C.,4.2.5.1.(6)
Anthony,P.,5.1.(19)
Aramati,V.S.,5.3.(6)
Arata,Y.,3.3.(27);3.4.(4);3.4.(5);3.4.(17);
  3.4.(22);4.1.1.(9); 4.1.1.(19);4.1.1.(31);
  4.1.1.(34);4.1.2.1.(28);4.3.4.(8);
  4.4.2.(22)
Arie,G.,5.1.(25)
Armstrong,S.,4.1.1.(21)
Arnot,R.S.,3.4.(10)
Arthurs,E.G.,5.3.(44)
Asaka,K.,4.1.3.(14)
Astrop,A.,4.2.2.(7)
Atkey,M.,4.2.2.(32)
Atthey,D.R.,3.3.(2)
Auvert,G.,5.3.(69);5.3.(70)
Averin,A.P.,3.2.(8)
Avramchenko,P.F.,4.1.1.(13);4.1.2.2.(9);
  4.1.3.(8);5.3.(16)

B
Babikova,Yu.F.,4.3.2.(8)
Bad'yanov,B.N.,1.4.(26)
Baeslack,W.A.,4.4.3.(8)

Bagdasarov,Yu.S.,4.4.3.(15)
Bailey,N.,4.3.4.(28)
Baker,G.,5.1.(35)
Bakowsky,L.,2.1.(32);5.1.(36)
Balbach,J.,2.1.(12);4.2.2.(15);4.2.2.(23);
  4.2.5.1.(8)
Balbi,M.,4.2.2.(21)
Balchin,N.C.,4.5.(29)
Baldasarov,Yu.S.,4.4.3.(16)
Banas,C.M.,2.1.(6);4.1.1.(3);4.1.2.1.(17);
  4.1.2.2.(5);4.1.3.(10);4.3.3.(3),4.4.3.(8)
Bandle,A.M.,4.5.(61)
Baranov,M.S.,4.1.2.1.(4);4.1.3.(6)
Barnaby,B.E.,4.1.2.3.(15)
Barnoski,M.K.,5.3.(2)
Barton,G.,4.4.2.(17)
Bashenko,V.V.,4.1.1.(20);4.1.1.(29);
  4.1.1. (30);4.1.2.2.(7)
Basov,N.G.,2.1.(29);4.1.1.(28)
Bass,M.,1.1.(4);1.4.(28);4.2.5.3.(2);
  4.2.5.3.(3)
Battista,A.D.,4.2.4.1.(3)
Bauer,F.R.,5.3.(14)
Bazan,M.,4.1.2.1.(29)
Be,C.A.,4.3.3.(17)
Beck,R.,1.4.(3)
Becker,W.,4.2.1.(12)
Bedrin,C.,4.3.1.(6)
Behnisch,H.,1.4.(50);1.4.(51)
Belanger,P.A.,3.2.(3)
Belen'kii,A.M.,4.1.2.2.(17)
Bell,T.,4.3.3.(15)
Bellows,G.,4.2.4.1.(5)
Belyeu,S.M.,2.1.(10)
Benedict,G.F.,5.1.(12)
Benzinger,M.,4.2.1.(9)
Bergmann,H.W.,4.3.3.(15);4.3.4.(23);
  4.4.2.(17)
Berloffa,E.H.,4.2.1.(13);4.2.3.(8)
Berrie,P.G.,4.2.4.3.(1)
Bertolotti,M.,5.3.(61)
Bever,E.,3.4.(6)
Beyer,E.,1.4.(41);3.2.(15);4.2.1.(21)
Biasio,S.D.,4.5.(51)
Billhardt,C.F.,2.3.(14);5.1.(31)
Birkett,F.N.,4.2.2.(44);4.2.4.3.(1)
Bishop,G.J.,4.3.5.(13)
Bitzel,H.,4.2.5.2.(4);4.2.5.2.(6)

Blake,A.,3.4.(14)
Blarasin,A.,4.1.2.2.(13)
Blom,G.M.,3.3.(14)
Blottner,F.G.,3.5.(12)
Bober,M.,3.3.(26)
Boccara,A.C.,4.1.2.2.(28)
Boccelli,V.E.,5.3.(10)
Bohme,O.,4.2.2.(6)
Bolin,S.R.,1.4.(2);1.4.(11);4.1.1.(4);
   4.1.2.1.(3); 4.1.3.(1);4.2.4.2.(5)
Bornstein,N.S.,4.3.1.(2)
Borrego,J.H.,4.1.2.3.(15)
Bousseau,M.,4.1.2.2.(32)
Bragard,A.,4.3.5.(4)
Bragg,M.J.,1.4.(49)
Breinan,E.M.,1.4.(13);4.1.1.(3);4.1.2.1.(17);
   4.1.2.2.(3);4.1.2.3.(3);4.1.2.3.(6);
   4.4.2.(14);4.4.3.(7);4.4.3.(13)
Brock,T.,4.1.1.(16)
Brouwer,E.,4.2.1.(24)
Bruck,G.J.,4.3.4.(25);4.4.3.(28)
Buness,G.,4.2.1.(3)
Burg,B.,2.2.(7);2.2.(9)
Burghardt,P.,4.4.2.(16)
Burris,M.K.,4.1.1.(14);4.1.2.2.(1)
Bushor,W.E.,1.4.(18);4.2.5.3.(5)
Butters,J.N.,5.4.(3)

C
Cai,M.,5.2.(17)
Califano,V.E.,5.3.(37)
Calvert,J.N.,4.1.2.2.(31)
Campello,M.,4.3.4.(26)
Cantello,M.,4.3.3.(4),4.3.4.(17)
Capello,G.,4.2.5.1.(7)
Carlson,D.W.,4.4.2.(33)
Caropreso,G.,5.2.(14)
Carroz,J.,2.2.(11)
Casey,H.,3.4.(1);4.1.2.1.(1)
Casey,K.W.,5.3.(21)
Cerri,W.,4.3.3.(17);4.3.4.(15)
Chabot,L.,4.5.(6)
Chang,U.I.,5.3.(21)
Charschan,S.S.,1.4.(28);4.5.(18)
Chatwin,C.R.,1.4.(30)
Chion,G.,5.3.(69)
Cielo,P.,2.1.(31)
Cingolani,A.,4.2.2.(42)
Chan,C.,3.5.(17)
Chande,T.,3.3.(20);3.5.(18)
Chang,D.U.,2.1.(18);3.5.(20)
Charissoux,C.4.1.2.2.(31)
Chen,G.,4.1.2.1.(20)
Chen,M.M.,3.5.(17)
Chene,J.J.,4.1.1.(6)
Chennat,J.C.,4.1.2.2.(29)
Chepurnov,V.I.,4.2.3.(15)
Cherkashin,A.P.,2.1.(11)
Chiaiese,V.C.,5.3.(45)

Cingolani,A.,4.2.4.2.(6)
Clark,J.N.,5.2.(1);5.2.(2)
Clarke,J.,4.2.1.(4)
Clement,P.,2.1.(39)
Clement,X.,4.3.5.(5)
Clough.R.B.,4.2.4.2.(8)
Cohen,M.G.,5.3.(28);5.3.(44)
Collin,M.,4.3.3.(13)
Com-Nougue,J.,4.3.2.(9);4.3.4.(22)
Cookson,J.,4.2.2.(45)
Cooper,E.B.,3.5.(13);4.2.5.1.(14)
Copley,S.M.,4.2.5.3.(2);4.2.5.3.(3);
   4.2.5.3.(8)
Coquerelle,G.,4.3.3.(13)
Corfe,A.G.,5.1.(20)
Coyle,R.J.,3.5.(11)
Cox,E.A.,4.5.(37);4.5.(44);4.5.(47);4.5.(55)
Crafer,R.C.,1.1.(6);4.1.2.1.(7);4.1.2.1.(9);
   4.1.2.1.(12);4.1.2.2.(18)
Crahay,J.,4.3.5.(4);4.3.5.(15)
Cusano,C.,4.3.3.(11)

D
Daene,K.,4.1.1.(11);4.1.2.1.(23)
Dale,B.W.,5.4.(5)
Dane,K.,5.3.(42)
Danielson,G.,4.5.(19)
Darchuk,J.M.,4.1.1.(33);4.2.1.(18)
Das,S.,4.3.3.(8)
Dasgupta,A.,4.4.2.(19)
Daurelio,G.,3.5.(5);4.2.2.(25);4.2.2.(35);
   4.2.2.(42)
David,S.A.,4.4.2.(6);4.4.2.(8);4.4.2.(9);
   4.4.2.(13);4.4.2.(19);4.4.2.(25);
   4.4.3.(18)
Davis,M.,3.1.(3);3.5.(14)
Davyov,V.A.,4.1.2.1.(18)
Dawes,C.J.,4.1.1.(26);4.1.2.2.(23);
   4.1.2.2.(24);4.1.2.2.(34);4.4.2.(21);
   5.3.(48)
De Pascale,O.,4.3.4.(13)
De Silva,G.M.S.,5.3.(31)
Debban,B.L.,4.1.2.3.(4)
DebRoy,T.,3.3.(21);3.3.(25);4.1.2.2.(26);
   4.4.2.(20);4.4.2.(29)
Decaux,J.-M.,1.4.(36)
Decker,I.,4.2.2.(41)
Dell'Erba,M.,3.4.(12);4.1.2.3.(13);4.2.2.(35)
Delle Piane,A.,2.2.(6)
Demuth,R.S.,2.1.(5)
Denney,P.E.,4.4.2.(23)
Dennis,R.B.,1.4.(48)
Dietz.J.,3.3.(33)
Dillard,P.A.,5.1.(22)
Dillio,A.,4.4.3.(23)
Dionoro,G.,4.4.2.(18)
Dixon,R.D.,3.4.(8);3.4.(16);3.4(19);3.4.(21)
Doig,A.T.,4.5.(12)

Donati,V.,3.3.(24)
Dorn,L.,4.1.3.(9)
Dowden,J.,3.1.(3);3.3.(28)
Doyle,D.J.,4.5.(60)
Drake,E.S.,5.4.(4)
Dullin,E.,2.3.(22)
Dumas,M.,3.1.(10)
Dumbadze,T.H.,4.3.3.(9)
Duncan,H.A.,3.4.(7);4.1.2.3.(13)
Dyatel,V.P.,3.1.(5);4.2.5.2.(7)
Dyer,P.E.,4.3.5.(13)

E
Eagar,T.W.,3.3.(10);4.1.2.3.(11)
Ebeid,S.J.,4.2.5.2.(3)
Eberhardt,G.,1.4.(33)
Eboo,M.,4.1.2.2.(6);4.3.3.(5)
Eckersley,J.S.,2.2.(5);4.2.2.(17);5.1.(14)
Edmunds,D.,4.5.(45)
Edson,D.A.,4.1.2.2.(18);4.1.2.2.(24)
Eiholzer,E.,4.3.3.(11)
ElSayad,M.M.,5.4.(10)
Elza,D,4.1.3.(15);4.5.(59)
Emery,M.H.,3.5.(6)
Engel,S.L.,2.1(3);4.1.2.1.(2);4.1.2.1.(5);
    4.1.2.1.(8)
Epstein,H.M.,3.4.(9)
Esposito,C.,3.5.(5);4.2.2.(42);4.3.4.(13)
Estill,W.B.,3.4.(11)
Evans,W.T.,5.4.(6)
Everley,M.,4.5.(56)

F
Fachinetti,J.L.,4.3.3.(13)
Fagan,W.F.,1.2.(4)
Fankhauser,F.,4.5.(1)
Fanning,W.J.,5.3.(51)
Fantini,V.,2.1.(28);2.3.(17)
Farmer,D.,4.5.(20)
Faulkner,G.E.,4.1.2.3.(8)
Fedorov,B.M.,3.5.(8);4.1.2.1.(13);
    4.4.2.(28);4.4.3.(20);4.4.3.(21)
Fedorov,V.G.,4.4.2.(2)
Feeley,J.T.,5.1.(29)
Fehrensen,H.,4.2.5.2.(13)
Feldman,M.,5.3.(8)
Ferrara,M.,4.2.2.(35);4.2.4.2.(6)
Ferraro,F.,4.3.4.(26)
Field,R.,4.2.2.(14)
Fieret,J.,2.1.(37)
Fiorini,O.,5.2.(19)
Fischer,R.,4.2.1.(15)
Flaum,M.,4.2.3.(12)
Fletcher,M.J.,4.5.(39)
Flick,F.F.,4.1.1.(21)
Forbes,N.,3.2.(17);5.3.(26)
Formisano,B.D.,3.4.(11)
Foulk,L.R.,5.4.(2)
Foulloy,L.,2.2.(7)

Forbes,N.,3.2.(17);5.3.(26)
Fournier,D.,4.1.2.2.(28)
Fraser,F.W.,4.1.2.3.(9);4.4.1.(4)
Friedrich,H.R.,5.3.(2)
Fritzsche,K.,3.5.(10)
Frye,R.W.,5.2.(16)
Fujioka,T.,1.4.(53)
Fujita,T.,4.2.5.2.(1)
Funkenbusch,A.W.,4.3.1.(2)

G
Galich,G.A.,5.3.(67)
Galsworthy,J.D.,4.4.2.(15)
Gandy,F.D.,5.3.(19)
Ganyushin,V.M.,4.1.2.2.(30)
Gavrilyuk,V.S.,4.4.1.(5)
Gay,P.,3.5.(9)
Geffroy,J.,2.2.(12)
Geiger,M.,4.2.5.2.(10)
Gensmer,W.,5.3.(22)
Georgalas,G.,2.1.(20);3.3.(18)
Gerbet,D.,4.3.4.(27)
Germani,G.F.,5.2.(19)
Gettleman,K.,5.1.(26)
Gilgenbach,R.M.,4.2.4.2.(7)
Gittos,M.F.,4.4.3.(22)
Gladkov,E.A.,3.1.(8);3.4.(15)
Gnanamuthu,D.S.,4.3.4.(20)
Goel,A.,4.2.2.(43)
Gol'tsova,V.P.,4.1.1.(5)
Gonzalez,J.V.,4.2.3.(7)
Goodman,D.S.,2.1.(21)
Gordon,J.R.,4.4.3.(22)
Gower,M.C.,4.3.5.(11)
Graham,R.E.,4.2.1.(23)
Gray,J.H.,5.2.(8)
Green,B.G.,1.4.(49)
Green,R.G.,4.2.4.2.(3)
Greenfield,M.A.,4.1.1.(3)
Greenslade,A.,4.2.5.2.(5)
Greenwald,L.E.,4.4.2.(14)
Greenwood,D.I.,4.3.5.(14)
Gregson,V.G.,1.4.(40);2.1.(40);4.2.2.(9);
    4.3.4.(3)
Grigor'ev,V.P.,4.3.2.(7)
Grigor'yants,A.G.,1.4.(22);3.3.(12);3.5.(4);
    4.1.2.1.(15);4.1.2.3.(12);4.4.3.(14)
Gruber,H.,3.5.(10)
Gudkov,V.K.,4.3.2.(7)
Gurvich,L.O.,4.3.4.(10)

H
Hagge,J.K.,5.3.(56)
Hahn,W.,4.2.2.(34)
Hakansson,K.,4.1.2.2.(12)
Hall,R.L.5.3.(33)
Hamilton,D.C.,4.2.4.1.(2)
Hanson,W.E.,4.2.3.(1)

Hardisty,F.B.,2.1.(35);2.3.(2);2.3.(8)
Hardock,G.,2.3.(20)
Harlen,F.,4.5.(27)
Haroutel,J.,4.1.3.(12);4.1.3.(13)
Harrison,T.W.,5.4.(1)
Hart,P.H.M.,5.1.(23)
Hartley,J.,2.3.(10);2.3.(21)
Hartung,P.,5.1.(29)
Hasson,D.F.,1.2.(2)
Hattori,N.,4.2.3.(5)
Hawkes,I.C.,4.3.4.(5)
Hayasaka,T.,4.1.3.(14)
Heglin,L.M.,5.1.(10)
Heinz,D.L.,3.3.(34)
Heiple,C.R.,4.4.2.(16)
Henderson,A.R.,4.5.(32)
Henderson,J.A.,5.3.(30);5.3.(32)
Hendrix,T.L.,5.3.(27)
Hendrixson,D.,2.2.(3)
Herbert,D.P.,4.2.2.(44)
Herbert,S.,5.1.(21)
Herbrich,H.,1.4.(21);4.2.1.(10);4.2.2.(28)
Herman,H.,1.4.(48)
Hernandez,J.,4.3.3.(14)
Herzinger,G.,3.2.(12)
Hill,M.,1.4.(17);4.1.2.2.(14);4.4.2.(12)
Hill,V.L.,5.1.(18)
Hodge,D.,4.2.5.2.(8)
Hoffman,M.,4.2.1.(5)
Hoffman,P.,2.1.(33)
Holt,R.D.,2.1.(13);2.1.(23)
Horne,D.M.,4.3.1.(3)
Horton,L.D.,4.2.4.2.(7)
Hoshinouchi,S.,4.1.2.2.(16);4.2.2.(19)
Huber,J.,4.2.2.(2);4.2.3.(6);5.1.(2)
Hugenschmidt,M.,3.1.(2)
Humphries,M.J.,4.3.1.(1)
Huntington,C.A.,4.1.2.3.(11)

I

Ianno,N.J.,3.3.(29)
Ikeda,M.,2.3.(16);3.3.(19);4.3.2.(4)
Inoue,K.,4.3.4.(8)
Irving,R.R.,5.1.(15)
Ishayama,H.,4.1.2.2.(27)
Itani,K.,2.1.(43)
Ivanov,V.V.,4.1.2.1.(24);4.4.3.(19)

J

Jakovlev,E.B.,3.5.(3)
James,D.J.,4.2.4.1.(2)
Janjua,M.S.,2.3.(4);2.3.(12)
Janssen,G.W.G.,5.3.(54)
Jeanioz,R.,3.3.(34)
Jefferson,T.B.,4.2.2.(4)
Jellison,J.L.,4.4.3.(1);4.4.3.(2)
Jerrard,H.G.,3.3.(2)
Jimenez,E.,4.1.2.3.(16)
Johnson,C.B.,3.1.(1)
Johnson,J.,4.2.5.1.(12)

Johnson,K.I.,4.1.2.2.(18);4.1.2.2.(24)
Johnson,R.,1.4.(17);1.4.(38);4.1.2.2.(14);
    5.3.(64)
Johnson,T.A.,2.2.(1)
Jon,M.C.,4.4.3.(4)
Jones,B.F.,4.4.2.(15)
Jones,M.G.,2.1.(20);3.3.(18);4.4.3.(6);
    5.3.(13);5.3.(41)
Jones,S.B.,4.1.1.(27)
Jones,T.A.,4.1.2.2.(11)
Jones,W.H.,3.3.(8);3.3.(11)
Jorgensen,M.,4.1.1.(15)
Jung,V.,5.3.(4)

K

Kalimullin,R.K.,4.3.3.(10)
Kamalu,J.N.,4.2.2.(8);4.2.2.(26)
Kapadia,P.,3.1.(3);3.5.(14)
Kaplan,R.A.,5.3.(28);5.3.(44)
Kashiwagi,T.,3.3.(9)
Katayama,S.,3.3.(32);4.3.4.(12);4.4.2.(31);
    4.4.2.(34);4.4.3.(26)
Kato,J.,4.2.4.1(7)
Kawai,Y.,4.1.2.2.(22)
Kawata,K.,2.1.(26)
Kaye,A.S.,2.1.(4)
Kear,B.H.,1.4.(10);1.4.(13);4.4.3.(7)
Kechemair,D.,2.2.(7);4.3.4.(27)
Keck,R.,1.4.(5)
Kelly,T.J.,4.4.3.(5);4.4.3.(10)
Kenney,G.,5.3.(11)
Kerrand,E.,4.3.2.(9);4.3.4.(22)
Kestenbaum,A.,5.3.(18)
Khan,P.A.A.,4.1.2.2.(26);4.2.2.(20)
Kiefling,L.,5.1.(33)
Kim,T.K.,5.2.(18)
King,P.P.,5.2.(6)
Kirillin,A.V.,4.2.5.3.(10)
Kirshnaswamy,H.N.,5.3.(10)
Kiszka,L.J.,5.3.(34)
Kiu,K.C.,5.3.(28)
Klemens,P.G.,3.3.(1)
Kobayashi,A.,1.4.(12)
Kobayashi,M.,4.2.2.(19)
Koechner,W.,2.1.(1);3.2.(1)
Kohls,J.B.,4.2.4.1.(5)
Kokhanovskii,V.D.,5.3.(24)
Kokora,A.,1.1.(1)
Kokosa,J.M.,4.5.(60)
Koledov,L.A.,3.3.(6)
Kong,R.L.,4.5.(41)
Konig,W.,4.2.5.2.(15)
Korlyakov,V.K.,4.1.1.(24)
Koschlig,M.,4.4.2.(17)
Koshy,P.,4.3.3.(7)
Kostrubiec,F.,3.3.(23)
Kosyrev,R.K.,4.2.1.(8)
Koura,M.M.,5.4.(10)
Kovalenko,V.S.,3.1.(5);4.2.5.1.(3);
    4.2.5.2.(7)

Kovalev,V.V.,4.1.2.2.(19);4.4.2.(27)
Kozhevnikov,Y.Y.,4.3.3.(10)
Kozlov,A.E.,5.1.(30)
Kramer,R.,3.2.(15)
Krauskopf,B.,4.2.5.3.(9)
Krutilla,M.A.,2.2.(19)
Kubel,E.J.,4.1.2.1.(27)
Kubo,L.Y.,5.3.(17)
Kubo,M.,2.1.(43)
Kudrayavtsev,E.P.,4.2.2.(13)
Kukreja,L.M..,4.2.3.(13);4.2.3.(16)
Kullen,J.,4.1.2.1.(10)
Kumehara,H.,4.2.4.1.(1);4.2.4.2.(4);
   4.2.5.1.(2);4.2.5.2.(9); 4.2.5.2.(11)
Kunieda,M.,2.3.(9);4.2.2.(31)
Kuvin,B.F.,4.1.1.(36)

L
Lagoutte,G.,4.1.3.(13)
Lakhtin,Yu.M.,4.3.4.(18)
Lamb,M.,4.3.1.(5)
Lamonde,G.2.1.(31)
Lampugnani,U.,4.1.2.1.(19)
Laney,J.C.,4.5.(26)
La Nier,M.E.,4.5.(14)
Laos,O.V.,1.4.(29);4.2.1.(19)
Largent,E.J.,4.5.(8)
La Roca,A.V.,3.3.(30)
Larsson,C.N.,4.2.5.2.(3)
Laude,L.,1.4.(44)
Laudel,A.,4.2.5.3.(4)
Laufer,G.,5.4.(9)
Lazarus,M.E.,5.2.(12)
Le,H.K.,4.3.1.(3)
Leather,J.A.,4.3.5.(3);5.3.(31)
Lebedev,V.K.,4.1.3.(7)
Lee,C.S.,4.2.2.(43)
Lee,S.,1.1.(7);4.3.3.(15)
Lee,S.-Y.,4.3.4.(30)
Lenz,E.,4.2.2.(22)
Lepore,M.,3.3.(15);4.2.1.(16);4.2.1.(17)
Lepoutre,F.,4.1.2.2.(28)
Levin,G.I.,4.1.2.1.(25)
Lewis,G.K.,3.4.(8);3.4.(16);3.4.(19);3.4.(21)
Li,L.J.,4.2.5.2.(14)
Lim,G.C.,3.2.(9)
Lindemanis,A.E.,4.3.3.(5)
Lingenfelter,R.C.,2.1.(22);2.1.(30)
Lippold,J.C.,4.4.2.(10)
Lison,R.,4.1.2.3.(10)
Litwin,A.,5.3.(49)
Liu,C.A.,4.3.1.(1);4.4.3.(18)
Loosen,P.,1.4.(47);3.2.(15)
Lotmar,W.,4.5.(1)
Lou,D.Y.,4.3.2.(1)
Luciani,P.Y.,4.1.2.2.(31)
Luft,A.,4.3.5.(12)
Lugara,M.,4.2.4.2.(6)
Lunn,D.J.,4.2.2.(10)

Luxon,J.T.,3.2.(11)
Lyman,J.,5.3.(29)
Lyman,O.R.,1.4.(14)
Lyon,T.L.,4.5.(53)

M
McCulloch,R.W.,5.2.(10)
McDonald,T.G.,3.2.(7)
Macfadyen,N.,3.2.(16)
McGuire,B.C.,3.3.(4)
Machulka,G.A.,5.3.(64)
Macintyre,R.M.,4.3.3.(2)
McKinlay,A.F.,4.5.(27)
Mackowiak,E.,5.3.(69)
McMillin,C.W.,4.2.3.(6)
Magnitskii,O.A.,4.1.2.1.(18)
Magrini,M.,4.3.3.(17)
Malachowski,M.J.,4.2.5.3.(12)
Mallory,M.B.,2.1.(23)
Mallow,A.,4.5.(6)
Maloney,E.T.,4.1.3.(1)
Manassero,G.,3.5.(9);4.2.2.(38)
Marchetti,R.,3.1.(6)
Marinsek,G.,4.2.5.1.(7)
Marshall,W.J.,4.5.(11)
Martyr,D.R.,5.1.(28)
Maruo,H.,3.3.(27);3.4.(22);4.1.2.1.(28)
Marx,W.,4.2.2.(2);4.2.5.3.,(1)
Masopust,O.T.,5.3.(9)
Masubuchi,K.,4.4.2.(3)
Masumoto,I.,4.1.2.2.(27)
Mather,J.C.,5.3.(56)
Mathur,A.K.,4.3.4.(19)
Matsumura,S.,4.3.4.(8)
Matsunaga,K.,4.4.3.(25)
Matsunawa,A.,3.3.(32);3.4.(13);4.4.2.(31);
   4.4.2.(34);4.4.3.(26)
Mayer,C.A.,4.1.2.1.(21)
Mayer,J.W.,5.3.(39)
Mazumder,J.,3.3.(20);3.4.(14);3.5.(2);
   3.5.(7);3.5.(17);3.5.(18);4.1.1.(25);
   4.1.2.1.(16);4.1.2.2.(15);4.1.2.3.(1);
   4.1.2.3.(7);4.2.5.2.(14);4.3.3.(11);
   4.3.3.(12);4.4.3.(11);4.4.3.(27)
Megaw,J.H.P.C.,1.4.(17);2.1.(4);
   4.1.2.2.(14);4.3.4.(4);4.4.2.(12);
   5.2.(1);5.2.(2)
Mehta,P.,3.5.(13);4.2.5.1.(14)
Meijer,J.,4.3.4.(29)
Melcher,R.L.,4.2.4.1.(6);5.3.(20)
Melonas,J.V.,4.1.2.1.(14)
Mendicino,P.A.,5.3.(55)
Merlin,J.,3.3.(33)
Metev,S.M.,4.3.2.(3)
Metzbower,E.A.,1.1.(3);1.2.(1);1.2.(5);
   4.1.1.(18);4.1.2.2.(20);4.1.2.3.(9);
   4.4.1.(1);4.4.1.(4);4.4.2.(1);

4.4.2.(23);4.4.2.(24);4.4.2.(32);
  4.4.3.(3);4.4.3.(9)
Migliorati,B.,4.2.1.(14)
Migliore,L.R.,4.1.1.(33);4.2.1.(18)
Miles,M.,4.5.(57)
Miller,F.R.,4.1.3.(2)
Miller,J.A.,2.3.(1)
Miller,R.,3.5.(13)
Minamida,K.,3.4.(3)
Mirkin,L.I.,4.3.2.(7)
Mironov,L.G.,2.1.(14)
Miska,K.H.,4.3.4.(1)
Mitani,M.,4.2.2.(39)
Mitchell,R.C.,5.2.(8)
Mitsuki,K.,5.4.(7)
Miura,H.,4.2.5.1.(15)
Miyabe,H.,3.3.(3)
Miyamoto,I.,3.3.(27);3.4.(22);4.1.1.(9);
  4.1.2.1.(28)
Miyata,T.,3.2.(14)
Miyazaki,T.,3.3.(3)
Mizutame,M.,5.2.(20)
Molain,P.A.,4.3.4.(19);4.4.2.(30)
Montanarini,M.,4.1.3.(3)
Montgomery,P.C.,5.4.(8)
Moon,D.W.,4.1.1.(37);4.4.1.(1);4.4.1.(4);
  4.4.2.(24);4.4.2.(32);4.4.3.(3);
  4.4.3.(9)
Moorhead,A.J.,5.2.(3);5.2.(7);5.2.(10);
  5.2.(13)
Moravskii,V.E.,5.3.(16)
Mordike,B.L.,1.4.(34)
Morgan,D.F.,2.1.(40)
Morgan-Warren,E.J.,4.1.2.3.(5)
Mori,M.,4.2.4.1.(1);4.2.4.2.(4);4.2.5.1.(2);
  4.2.5.2.(9);4.2.5.2.(11)
Moriyasu,M.,2.1.(36)
Moss,C.E.,4.5.(34)
Mottier,F.M.,2.1.(8)
Mucci,J.,4.4.2.(5)
Muniz,J.M.,5.3.(37)
Murakawa,M.,4.2.2.(37)
Murchie,J.R.,4.1.2.3.(8)
Murray,W.E.,4.5.(14)
Muthukrishnan,S.,4.1.1.(32)

N
Nagai,H.,1.4.(46)
Nagano,Y.,4.2.4.1.(4)
Nakagara,S.,4.2.5.2.(1)
Nakagawa,T.,4.2.2.(31);4.2.5.1.(11)
Nakamura,E.,2.3.(19)
Nanba,H.,3.2.(14)
Nannetti,C.A.,4.3.4.(26)
Narayan,J.,4.4.1.(3)
Naugler,T.W.,1.4.(4)
Neiheisel,G.L.,4.3.1.(4)
Newman,R.A.A.,5.1.(4)
Nicolas,S.,4.3.5.(16)

Nielsen,S.E.,4.2.2.(49);4.2.3.(10)
Nikolaev,G.A.,1.4.(22)
Nilsson,K.,2.1.(15)
Nishio,R.,3.4.(4)
Nishiyama,N.,4.1.2.2.(10)
Norrman,S.,5.3.(60)
Notenboom,G.J.A.M.,5.3.(52)
Novozhilov,N.M.,4.1.2.2.(19);4.4.2.(27)
Nowack,R.,3.2.(19)
Nunes,A.C.,3.1.(7)
Nurminen,J.I.,4.3.4.(25)

O
Oakley,P.J.,1.4.(19);1.4.(52);4.1.1.(26);
  4.3.4.(11);4.3.4.(28)
Oda,T.,3.4.(4);3.4.(5);3.4.(17);4.1.1.(31);
  4.4.2.(22)
Ofer,V.I.,4.1.2.2.(21)
Oh,J.E.,3.3.(29)
Ohlschlager,E.,4.1.3.(9)
Ohmine,M.,2.3.(19)
Ohmura,E.,3.5.(15)
Olsen,F.O.,2.1(16)
O'Neal,J.E.,4.4.3.(12)
Ono,A.,2.1.(7);4.1.1.(10)
Osada,G.,4.2.2.(43)
Osterink,L.,1.4.(16)
Ostretsov,Yu.N.,5.3.(12)
Ott,G.,4.5.(48)
Otten,R.,4.2.5.1.(14)

P
Pal,G.S.,4.2.3.(14)
Pandarese,F.,5.2.(19)
Papazoglou,V.J.,4.4.2.(3)
Parker,R.D.,5.3.(5)
Parks,P.F.,5.3.(1);5.3.(58)
Parrini,C.,4.1.2.2.(5)
Parsons,G.H.,4.3.4.(6)
Pauleau,Y.,5.3.(70)
Peebles,H.C.,3.4.(18)
Peerey,R.S.,1.1.(2)
Penco,E.,3.1.(6)
Peng,Y.C.J.,1.4.(16)
Perry,F.,4.1.1.(21)
Petesch,B.,4.1.2.3.(18)
Phillips,R.H.,4.4.2.(32)
Pippen,C.A.,4.2.5.2.(2)
Plankenhorn,D.J.,5.1.(18)
Poate,J.M.,5.3.(39)
Polad,M.,2.3.(13)
Polk,D.H.,4.2.4.3.(2);5.2.(16)
Pope,L.E.,3.2.(7);4.4.3.(2)
Popova,L.V.,4.2.1.(23)
Pothoven,F.,4.1.3.(11)
Powell,J.,4.2.2.(36);4.2.2.(44);4.2.2.(48)
Price,C.,4.5.(40)

Q

Querry,M.,4.3.1.(6)
Quinlan,J.C.,2.2.(2)

R

Raevich,V.K.,4.2.3.(2)
Raffel,J.I.,5.3.(46);3.5.(68)
Rajagopal,S.,2.3.(3);5.1.(18)
Rajaram,S.,3.5.(11)
Ramos,T.J.,2.1.(22);2.1.(30);4.1.2.3.(8)
Ramous,E.,4.3.3.(16);4.3.3.(17);4.3.4.(15)
Rand,M.J.,5.3.(25)
Rasher,A.,4.2.3.(6)
Rathmill,K.,2.3.(4)
Ravich,G.N.,5.1.(13)
Raybourn,M.S.,4.5.(41)
Ready,J.F.,1.4.(25)
Ream,S.L.,5.2.(11)
Reed,W.E.,1.3.(1);1.3.(2)
Reid,R.D.,5.2.(5)
Ricciardi,G.,4.3.4.(7)
Rioux,M.,3.2.(3)
Ripper,G.,3.2.(12)
Robert.H.,3.3.(7)
Rockwell,R.J.,4.5.(13);4.5.(34)
Rogalsk,W.J.,5.1.(9)
Rogell,M.L.,5.2.(8)
Roos,S.-O.,2.2.(10);4.2.4.2.(2)
Roper,J.R.,4.4.2.(16)
Ross,I.E.,2.1.(25)
Roth,M.,4.3.4.(17)
Rothe,R.,4.2.2.(16);4.2.2.(29)
Rourke,M.,1.4.(31)
Roy,S.,4.2.1.(2)
Ruge,J.,4.1.2.1.(20)
Russell,J.D.,4.1.1.(27)
Russell,R.M.,5.4.(6)
Russo,A.J.,3.3.(22);3.5.(19);4.1.3.(17)
Ryba,E.R.,4.1.3.(4)
Rykalin,N.,1.1.(1);3.3.(5);3.3.(13)

S

Safonov,A.N.,4.3.4.(16)
Saifi,M.A.,4.1.1.(2);4.3.5.(1);5.3.(3);5.3.(8)
Salvetti,G.,3.1.(6)
Salzer,T.E.,5.3.(40)
Sanderson,R.J.,1.1.(5)
Sano,R.,3.2.(18)
Sarady,I.,2.1.(15)
Sasaki,H.,4.1.2.2.(10)
Sasnett,M.W.,3.3.(31)
Sauer,M.,5.3.(4)
Saunders,R.J.,1.4.(37)
Sayegh,G.,4.1.1.(38)
Scarff,P.L.,5.3.(34)
Schaefer,R.J.,4.2.4.2.(8)
Schaffer,G.,4.2.2.(1);5.1.(11)
Schekulin,K.,4.2.1.(11)
Schellhorn,M.,3.2.(19)
Scheuermann,W.,4.2.1.(6)

Schmitt,A.J.,3.3.(17)
Schock,W.,2.1.(41)
Schreiber,P.",4.5.(48)
Schuocker,D.,3.5.(16);4.2.3.(11)
Schwartz,M.M.,4.1.1.(12)
Schwarz,M.,2.3.(15)
Schwob,H.P.,5.1.(24)
Scott,B.F.,1.4.(1);1.4.(30)
Scott,M.H.,4.4.3.(22)
Seaman,F.D.,2.3.(3);3.2.(4);4.1.2.1(6);
　4.1.2.1.(11);4.1.2.2.(2);5.1.(8)
Seiler,P.,5.3.(35);5.3.(59)
Semiletova,E.F.,4.3.3.(9)
Senin,A.,4.3.4.(26)
Sepold,G.,4.2.2.(29);4.4.2.(26)
Seretsky,J.,4.1.3.(4)
Shaber,E.L.,4.1.2.3.(4)
Shachrai,A.,4.2.2.(22)
Shankar,V.S.,4.3.4.(20)
Sharp,C.M.,1.4.(7);1.4.(24);3.2.(5)
Sharp,M.,1.4.(15)
Shewell,J.R.,4.1.1.(7)
Shimada,W.,4.2.2.(19)
Shimo,C.,4.2.4.2.(9)
Shiner,W.H.,4.2.4.1.(3)
Shinmi,A.,4.1.2.2.(33)
Shinoda,T.,4.1.2.2.(27)
Shinohara,K.,3.3.(19)
Shiroki,K.-I.,4.2.5.1.(15)
Shovkoplyas,V.M.,3.2.(6);4.1.1.(13);
　4.1.3.(8)
Sibayama,K.,2.1.(43)
Sidorowicz,K.,3.5.(1)
Signamarcheix.J.-M.4.1.2.2.(32)
Silberglitt,R.S.,4.3.1.(3)
Silva,G.,4.2.2.(21)
Silva,R.M.,4.2.1.(1)
Silva,S.D.,4.3.5.(3)
Simpson,G.,5.3.(53)
Singer,J.,3.3.(26)
Singh,J.,4.3.3.(12);4.4.3.(27)
Skripchenko,A.I.,3.2.(10);4.1.1.(30)
Slack,R.B.,1.4.(8)
Sliney,D.H.,4.5.(3);4.5.(5);4.5.(13);4.5.(16)
Smart,D.V.,5.3.(49)
Smart,R.,4.5.(17)
Smeggil,J.G.,4.3.1.(2)
Smith,J.E.,4.3.4.(25)
Smith,R.,2.3.(6)
Smith,W.H.,4.4.2.(13)
Smoluk,G.R.,1.4.(39)
Snyder,D.G.,4.2.3.(14)
Snow,D.B.,4.1.2.3.(3);4.4.2.(7);4.4.2.(14);
　4.4.3.(13)
Sobol,E.N.,4.3.4.(10)
Sooy,W.R.,2.1.(1)
Spalding,I.J.,2.1.(17)
Steen,W.M.,1.4.(20);1.4.(24);2.1.(38);
　3.2.(9);3.5.(2);3.5.(7);4.1.1.(17);
　4.1.2.2.(6);4.1.2.2.(15);4.1.2.3.(1);

4.1.2.3.(7);4.2.1.(4);4.2.2.(8);
 4.2.2.(26);4.3.1.(5);4.3.3.(1);
 4.3.3.(6);4.4.3.(11)
Steffen,J.,4.1.3.(3)
Stenow,A.,4.5.(21)
Stern,G.,4.3.4.(14)
Stevenson,P.,4.1.1.(39)
Stickney,B.,5.3.(64)
Stockham,N.R.,5.3.(48)
Stoop,J.,4.4.2.(1)
Strychor,R.,4.4.2.(24)
Stumer,E.,4.2.4.2.(1)
Suess,M.J.,4.5.(24)
Sugihara,K.,4.2.5.2.(1)
Suminov,I.V.,5.3.(57)
Sumiya,M.,2.1.(26)
Surkov,A.V.,3.2.(10);4.1.1.(30);
 4.2.2.2.(19);4.4.2.(27)
Sutugin,A.G.,4.2.1.(22)
Swenson,E.J.,5.3.(65)
Szadkowski,D.,4.5.(25)

T
Tabata,N.,2.3.(18)
Takahashi,K.,3.2.(13)
Takaoka,T.,4.2.5.1.(1)
Takeda,T.,4.3.3.(6)
Tal,Y.,4.2.2.(22)
Tamaschke,W.,4.2.2.(12)
Tereshkow,H.,5.1.(5)
Terrell,N.E.,5.1.(17)
Teske,K.,4.2.2.(47);4.4.2.(26)
Teti,R.,4.4.2.(18)
Thomassen,F.B.,2.1.(16)
Thompson,E.R.,1.4.(13)
Thorn,R.,1.4.(45)
Tiffany,W.B.,4.3.5.(10)
Tight,T.,2.2.(13)
Tikhomirov,A.V.,1.4.(32);4.2.2.(3);4.2.2.(13)
Toenshoff,H.,4.2.2.(15)
Tomlinson,J.,5.3.(15)
Tonneau,D.,5.3.(70)
Tonshoff,H.K.,2.1.(12);4.2.2.(23);4.2.5.1.(8)
Tosch,R.,3.5.(10)
Townsend,A.B.,4.1.2.3.(2)
Townsend,T.A.,2.1.(9)
Trefilova,B.,5.3.(7)
Tremblay,R.,3.2.(3)
Trenholme,J.B.,3.2.(2)
Tsuboi,J.,4.1.2.2.(10)
Tsushima,K.,4.2.5.1.(15)
Turned,P.W.,4.1.2.3.(2)
Tuz,Z.L.,1.4.(43)
Tyrer,J.,5.4.(8)

U
Ueda,K.,2.1.(26)
Uetz,H.,2.3.(20)
Uglov,A.,1.1.(1);3.3.(5);3.3.(13);4.1.2.2.(21)

Ulrich,O.E.,4.2.4.2.(7)
Un-Chul Paek,4.3.5.(1)
Urbach,F.,4.5.(2)

V
Vaccari,J.A.,4.2.5.3.(11)
Vahaviolos,S.J.,4.1.1.(2)
Vahl,J.,5.1.(7)
Van Benthem,H.,5.1.(7)
Van Cleave,R.A.,4.2.3.(3);4.2.3.(9)
Vanderwert,T.L.,4.1.3.(16);4.2.2.(27);
 4.2.5.1.(13);4.2.5.3.(7);4.3.4.(9);
 5.1.(37)
Van Dijk,M.H.H.,4.2.1.(24)
Vanschen,W.,4.2.2.(5)
Van Scoy,R.L.,2.1.(27)
Velichko,O.A.,4.1.1.(8);4.1.1.(13);4.1.3.(8);
 5.3.(16)
Velikhov,E.P.,1.4.(35)
Velikikh,V.S.,4.3.4.(24)
Vendramini,A.,4.3.4.(15)
Verkhoturov,A.D.,(7)
Veverka,D.B.,5.2.(11)
Vitek,J.M.,4.4.2.(6);4.4.2.(8);4.4.2.(9);
 4.4.2.(13);4.4.2.(19);4.4.2.(25)
Vito,A.D.,4.1.2.2.(5)
Volobuev,Yu.V.,3.3.(16);4.1.1.(22)
von Felixburger,F.,4.5.(25)
von Gutfeld,R.J.,4.3.2.(2)
Voshchinskii,M.L.,4.1.3.(5)
Vulpetti,A.T.,4.5.(50)

W
Wadley,H.N.G.,4.2.4.2.(8)
Wadsworth,J.,4.4.3.(17)
Wakagawa,T.,2.3.(9)
Walker,R.W.,4.2.5.1.(10)
Wallace,R.G.,4.2.5.3.(3);4.2.5.3.(6)
Wang,H.-P.,4.4.3.(6)
Ward,B.A.,2.1.(34);2.1.(37)
Ward,G.R.,4.5.(61)
Warnecke,H.-J.,2.3.(20)
Wasko,J.H.,4.2.5.1.(4)
Watanabe,T.,4.2.4.2.(9)
Waters,R.L.,4.1.3.(11);5.1.(13)
Watson,M.N.,4.1.1.(26);4.2.1.(20);4.4.2.(21)
Wautelet,M.,1.4.(44)
Webb,R.,1.4.(28)
Weber,S.,4.2.1.(3)
Weedon,T.M.W.,1.4.(9);1.4.(27);5.1.(27)
Weerasinghe,V.M.,2.1.(38);4.3.3.(1)
Weeter,L.A.,4.4.3.(24)
Weick,J.,4.2.1.(9);4.2.2.(40)
Weil,B.S.,5.2.(15)
Wellendorf,K.,4.2.2.(33)
Welty,R.K.,5.2.(4);5.2.(5)
Werth,D.L.,4.1.2.3.(17)
West,B.,5.1.(32)
West,D.R.F.,4.3.1.(5);4.3.3.(6)

Westkott,D.,1.4.(21)
White,C.W.,1.1.(2);5.3.(50)
Whitlock,R.R.,3.1.(9)
Whittaker,J.W.,4.1.1.(35)
Wilcher,F.E.,4.5.(43)
Wildish,M.,2.3.(11);4.3.5.(7)
Wilhelm,H.,1.4.(5);4.2.2.(30)
Willgoss,R.A.,5.2.(1);5.2.(2)
Williams,V.A.,1.4.(42)
Willis,J.B.,4.3.5.(8)
Winburn,D.C.,4.5.(42)
Winship,J.T.,2.3.(5)
Wissmeier,H.-J.,4.2.5.2.(10)
Wittkopp,C.,2.3.(14);5.1.(31)
Witzmann,J.,4.2.1.(13);4.2.3.(8)
Wolbarsht,M.,4.5.(16)
Wolf,G.,4.1.2.1.(26)
Wollermann-Windgasse,R.,2.1.(42);4.2.2.(40)
Wood,R.F.,5.3.(50)
Wright,J.K.,1.4.(9)
Wuthrich,R.,5.3.(36)

Y
Yakushin,B.F.,4.4.3.(21)
Yamada,A.,3.3.(19)
Yamada,K.,5.3.(38)
Yamaguchi,H.,5.3.(62)
Yeack,C.E.,4.2.4.1.(6)
Yoshioka,S.,3.3.(3)
Young,C.G.,1.3.(3)
Young,M.,4.3.4.(23)
Young,R.T.,5.3.(50)
Yuan,S.F.,4.3.1.(6)

Z
Zaiguang,L,4.3.4.(21)
Zechmeister,H.,4.2.5.1.(9)
Zemskov,K.I.,4.3.2(6)
Zhuk,F.I.,5.3.(47)